Permaculture Design

by Aranya

A step-by-step guide

Permanent Publications

Published by
Permanent Publications
Hyden House Ltd
13 Clovelly Road
Portsmouth
PO4 8DL
United Kingdom
Tel: 01730 776 582
Email: enquiries@permaculture.co.uk
Web: www.permanentpublications.co.uk

Distributed in North America by
Chelsea Green Publishing Company, PO Box 428, White River Junction, VT 05001 USA
www.chelseagreen.com

Distributed in Australia by
Peribo Pty Limited, 58 Beaumont Road, Mt Kuring-Gai, NSW 2080 Australia
https://peribo.com.au

First published 2012
Rreprinted 2013, 2014, 2015, 2016, 2017, 2018, 2019, 2020, 2021, 2022, 2024

Designed by Aranya

Printed in the UK by Bell & Bain, Thornliebank, Glasgow

MIX
Paper | Supporting responsible forestry
FSC
www.fsc.org
FSC® C007785

This product is made of material from well-managed FSC®-certified forests and from recycled materials and other controlled sources.

The Forest Stewardship Council ® (FSC) is a non-profit international organisation established to promote the responsible management of the world's forests. Products carrying the FSC label are independently certified to assure consumers that they come from forests that are managed to meet the social, economic and ecological needs of present and future generations.

British Library Cataloguing-in-Publication Data
A catalogue record for this book is available from the British Library

ISBN 978 1 85623 091 9

Contents

PART THREE ~ Beyond Land Based Design

Acknowledgements

Offering my deepest thanks seems simplest in chronological order, so I shall start with my parents Rod and Val who've always been there for me, even when they must have thought I'd lost the plot. I love you both dearly. Next, to Bill Mollison and David Holmgren, I thank you for pulling together the best of humanity's vision and calling it permaculture. Like their own, this book is dedicated to Gaia and all those who sail in her...

More specifically, of the early pioneers, Tanya was the first to introduce me to the 'P' word all those years ago. Steve Charter made sure I couldn't escape by inviting Stephen Nutt to teach a design course at our shared house (I am most indebted that you ran the course for just four of us). Tony Wright added a whole new angle to the big picture I was getting. I thank you all. My experiential journey then began in the Emerald Isle at Tiaia, where I learned the gifts of barefooting and many other things besides. Giri and Sagara, you challenged me deeply but showed me so much in that single year. I learned from implementing there what I read in books by Robert Hart, Patrick Whitefield, Ken Fern, Masanobu Fukuoka and Robert Kourik.

On my return to England I made gardens and connected into the network, learning much from many peers and mentors in those early days – Julia, Sandra, Jane, Simon, George, Pat, Sue, Ann and Phil – you've all played an important part. Andy, Chris and Looby, thanks for having the foresight to set up Designed Visions together and for all you have taught me along the way. A massive thank you to Andy Goldring for all he's done at the Permaculture Association. Hannah, Peter, Klaudia, Mel B, and everyone else I've taught with, I love the journey. Mel C, thank you for freeing up my time to write.

On to the book itself; a great big thank you Tim and Maddy for sticking your necks out and publishing my first book. Gary and Katy, thank you both for your valuable feedback on the draft.

Finally, Jules. Without all your proofreading and looking after mundane tasks to free up my time, I certainly wouldn't have got this far. Yet the greatest gift you give is your belief in me and your love. Thank you so much for being in my life and making the journey so much fun. May it be a long one. xxx

Foreword

by Patrick Whitefield

Some people keep their tools in a higgledy-piggledy pile, chipped, blunt and uncared-for. Others keep them in perfect order in a neat toolbox, each one in its place, oiled, sharpened and ready to go. Aranya is one of the latter and this book is his tool box. Open its pages and you can put your finger right on the design tool you need, from triangulation to the enticingly-named desire lines. It's all here.

But the tool metaphor can be taken too far. Permaculture design is not a random collection of good ideas, it's a process. It starts with the receptive stage of observing and listening, goes through the creative stage of making design decisions and leads on to implementation on the ground and to reflection on whether the odd tweak here or there might not be appropriate. Aranya takes you through the process with the same chisel-sharp clarity as he describes the individual design tools.

For most people taking the permaculture design course is a major life event. It opens the door on a new way of looking at the world, a new way of being in the world. It affects you on so many levels that, however well you took notes, inevitably some of the information passes you by. That's why I wrote *The Earth Care Manual* and that's why Aranya wrote the book you have in your hands.

Patrick Whitefield
Permaculture designer, teacher and author
Glastonbury 2012

Preface

This is not just another permaculture book. There are already plenty of excellent texts[†] on this subject explaining the many permaculture design principles, strategies and techniques that can help make the world a better place for us all to live in. This guide is a little different. It's been written to help those with a basic grasp of permaculture to gain more clarity about the overall design process.

In my role as a teacher, I meet many students who've done an introductory course or read a book or two and got a basic understanding of the key concepts, yet are still unclear how to put them all together. As integration is the key to good design, this guide sets out to make sense of all this and create a useful pattern that you'll be able to apply to your own design work.

This guide started out as a series of worksheets, created to assist students on my weekend-based permaculture design courses. I'd noticed that with gaps of several weeks between course weekends, students were often forgetting important details. I wanted to provide them with a clear pathway that they could follow while going through their design practice at home.

After introducing the new worksheets, student's design work greatly improved, so as an experiment I started giving them to the design groups on my residential courses too. Feedback was again very positive. Then some students suggested that I should publish them as a book.

So here it is. I hope that this guide helps to clarify the design process for you, and gives you the confidence to go out and gain the experience that will ultimately make you the accomplished permaculture designer that you wish to be.

Enjoy the journey!

† *See my recommended reading list in the appendices.*

How to use this guide

The first thing to say about this guide is that it's intended to spend time out and about with you, to be your companion on your permaculture design projects. For this reason I've made it small enough to fit into a bag or large pocket. In the tradition of permaculture multi-functionality, I've written it so it can be read either cover to cover, or simply dipped into from time to time for fresh ideas. That said, I do recommend you read it all the way through at least once, in order to get a full sense of the designer's journey, then work with it to put everything you've already learned about permaculture into the context of the design process. I've included a lot of detail for experienced designers, but if you're new to this don't feel you have to do every single thing, just follow where your interest takes you. There's plenty of time to learn the finer points.

When writing the guide, I set out to focus purely on the design process itself, leaving details of techniques to be gleaned from any of the other books already written about permaculture. However, I ended up including a part at the beginning that introduces the world of systems and patterns. I've done this as I feel that a basic understanding of how the world around us works is fundamental to being able to design anything within it, and that these things aren't often taught elsewhere.

The book has three main parts, each divided into sections. Each section has been given its own icon, printed on page corners to help you find them easily. Each finishes with a summary of the main points and most also with a flowchart for those who prefer it. Throughout the text I've highlighted ***principles and directives*** and ***design tools and techniques*** like this when I first mention them. The appendices include a glossary of less common terms and a collection of other resources, some of which I have provided for you as online downloads, identified in the text by this symbol: ¤. I hope they support your learning journey and help you to become a better designer.

So just to remind you – while it would of course look very nice sat on a bookshelf, this guide *longs to live a little* and to get a bit creased around the edges. Please make it your friend. ☺

Introduction

So, what attracted you to permaculture? If you're like most of the people I meet on courses, you'll have felt you'd discovered something that can help you make a difference in the world.

Perhaps you visited an inspirational project and heard the word mentioned, or someone gave you a book and you got excited about better ways of growing both food and community?

While permaculture attracts people of all ages and from many walks of life, the thing I've found most common to all of them is an urgent desire to do something about the mess we've been making of the world. I feel that urgency myself, yet ironically through permaculture, I've discovered that the tortoise really can be quicker than the hare. Or at least end up working less hard to get there.

I've learned this the hard way, by for instance my impatience to make gardens. I once laid out a few pallets around a mobile home where some decking was planned, so I knew where to start making my mulch beds. Only later did I start thinking about all the different uses I could make of such a deck and that two pallets deep was a much more sensible size. I spent a lot of time rescuing plants from the shade under that structure while I was building it.

So ironically, one of the most important things I do these days is to *slow people down*. To make sure that they take the time to really notice the landscape with which they are interacting. To consider all the options available to them, not just those that come straight to mind (perhaps because they saw them on TV last night). To think long-term about how their own needs and the landscape around them could change (*and it will*).

We need to take action, but let's make sure we plan that action well, so we don't regret it later. The good news is that, with permaculture in your toolbox, you'll have most of what you need to do a fine job. Let's open the lid and look inside...

You're already a designer

Every time you make a meal, or rearrange your furniture, you're designing. You might not have considered this, because you're doing analysis in your head and drawing from your previous experience, but you're still designing. You're *already* a designer! What the permaculture design process and this book in turn sets out to do though, is to give you a framework and a good collection of thinking tools to make your process both clearer and more consistently successful.

Whenever you make that meal or move that furniture around, it's a relatively quick process with immediate effect. Both try out ideas in the real world; if the meal isn't so tasty, there'll soon be another opportunity to do better. If the furniture doesn't look quite right, you can easily move it again. Neither is serious if it doesn't go quite to plan, as relatively little energy is invested in the process.

However, when you start designing a garden or smallholding, a building, or a whole new livelihood, you have much more at stake. Once you've built your house or made your garden, once you've handed in your notice at work, any mistakes are much harder to correct. Bill Mollison calls these 'Type 1' errors – those we regret making every day afterwards because they make it so hard to get anything else right.

Life experience may be a fairly reliable guide in our interactions with small day-to-day stuff (it's safe to eat *this* or to do *that*), but when we start interacting with the world in a bigger way, we begin to get unexpected results. These larger systems don't always behave as we expect them to, so it's useful to have some design tools to help us anticipate the things that may go wrong and to design in some 'safety nets'.

Permaculture provides us with an evolving toolkit,[†] to help us successfully develop complex designs. The process described in this guide is not '*the only way*'; rather it is one strategy, which I hope you'll find useful as a basis for your own.

† *Each of us who interact with permaculture add to its scope, evolving it further.*

PART ONE
~
Preparations

Observation Skills

Nature provides the best examples we have of complex systems that are not just abundant, but resilient and self-balancing too. Indeed we are completely surrounded by examples of excellent practice – including our own bodies. Yet it's clear from the imbalances that we are now creating in natural systems, that our recent human actions haven't been quite so well designed. We know we can do better, and the feedback that we are receiving is part of our education.

The fast pace of modern life means that while we may look, we don't always see. Our attention span gets shorter as we reach information overload, and yet the things we often notice the least are the very things keeping us alive, day in, day out. To turn things around, we have to learn to see again, which is why permaculture urges us to become better observers. Starhawk shares nine different ways for us to do this in her book, *The Earth Path* and nature awareness training also has many such gifts to share. Observation skills were vital for our ancestors just to stay alive and despite our seemingly cosy modern lives, that may not be so far from the truth for us today either. The dangers may be less apparent to us, but unheeded they could be just as deadly.

Thankfully, many of us are already being good lookouts for humanity, pointing out the potential dangers of climate change and *peak oil*† to others. As astute observers of what has been going wrong, we're also part of the way to creating solutions. We just need to learn from our mistakes and see where we could make better choices in the future. Good observation is an important skill for us to have in gathering information that enables us to create those better designs, and we'll need it to monitor the effectiveness of our subsequent creations too.

† *The point when the maximum rate of global petroleum extraction is reached and shortages become an issue. Many oil experts feel we may be at that point already.*

Patterns

One of the first things we notice when we observe natural ecosystems is that certain ***patterns*** keep appearing, in many situations and at varying scales. These patterns occur in both time and space, and while the former determine our routines, the latter are often only considered for their beauty. However, nature's most common patterns have evolved over many millennia† as being the *most effective* for survival. While conditions on the surface of the Earth have changed considerably over time,‡ life has always managed to adapt in order to survive and thrive here. So our challenge as designers is to identify what each of these patterns excels at doing and to apply them where appropriate in our designs.

Patterns in space

The study of patterns and their successful application in design is a fascinating and detailed subject much beyond the scope of this guide. That said, there are some key principles that, once understood, can help us to use them effectively in our designs. Firstly, patterns occur at the ***edge*** between two different media or systems. So the branching fractal form of a broccoli head is simply the edge between the plant and the atmosphere, the waves on the ocean the place at which the air mixes with the water. Resources are exchanged across these edges; needs are met and waste products eliminated, so by increasing surface area, nature increases the efficiency of this interaction and ultimately the size and health of the organism. Hence we find that many of the most common patterns we see around us in nature (branching, waves, spirals, webs and so on), all have extensive 'edge'. Of course the ultimate edge on this beautiful planet is the one between the earth and the atmosphere where almost everything lives, and that's because this is where all the key requirements for life occur together.

† *Nature has done 3,500,000,000 years of Research & Development!*

‡ *Levels of Oxygen in the atmosphere were once much lower and the temperature of the sun also considerably cooler, requiring higher levels of greenhouse gases.*

Think about what soil actually is – minerals from the earth, mixing with gases from the atmosphere, and water from both regions. This is why healthy soil has to have a good structure, in other words *lots of edge*, in order to support life! Life in turn speeds up the interaction between the earth and the atmosphere; plants and trees in particular, considerably increasing the surface area of exchange between these two media through the branching of their top growth and root systems. Thus, these *translators*[1] keep energy and resources such as sunlight, minerals, water and gases on the move across this familiar edge; harvesting sufficient to grow and make themselves ever more effective at performing this service, while also building soil as a by-product, which in turn supports more life.

Of course there are times too when minimising edge (to reduce exchange) is advantageous. When a woodlouse curls into a ball for protection, or a cat curls up to retain body heat, we see this occurring. As well as having the least edge, spheres are also the most stable form, and yet nature is always looking for ways to increase that edge while maintaining a dynamic stability. A cell will stretch to increase its surface area and upon reaching a point of instability will divide into two stable forms again, but now with a greater combined surface area. This process occurs over and over again, in a pulsing form. Our own bodies are only prevented from returning to a stable spherical blob by our investment in a skeleton to keep us elongated. Many other life forms develop similar structures to increase their surface area.

Even the surface of the Earth, though close to spherical in form, is constantly being disturbed by tectonic activity pushing up mountain ranges and by wind driving up waves on the oceans. Those great mountain ranges are gradually ground down by the forces of nature, into smaller and smaller pieces, with each fracture increasing the edge upon which life can take hold. So contrary to our fear of things falling apart, this breaking down process actually provides more opportunities for life.

We've been taught to fear entropy,† but without this process there'd be no raw materials for life to create itself anew with. The 'Lobular' pattern that results from weathering is what makes a well-structured soil fertile and such an effective cleaner of water overloaded with faeces; a pattern we mimic in using gravel beds to do this job in sewage treatment works, gravel providing a substantial surface area on which bacteria that break down sewage can live. We find the same pattern again at work in our own colons, where bacteria pre-digest our food for us through their huge collective surface area.‡ Don't forget that decomposers perform the vital recycling role within ecosystems.

Beach stones exhibit the lobular pattern

We see many patterns of beneficial interaction around us in nature. Our own bodies are of course prime examples of the value of successful ***co-operative relationships*** between cells. In addition to an abundance of multi-cellular organisms, nature abounds with symbioses – close and often long-term interactions between different biological species – and we can identify patterns in the ways that they interact with each other. Even apparently parasitic relationships reveal their mutualisms upon closer inspection. Ivy is often blamed for killing trees and cut down to stop this occurring. But how can such a relatively small plant ever overwhelm a large healthy tree? Ivy provides important wildlife habitat and cutting it causes more harm than good. Careful study reveals that it only overwhelms trees that are already dying, the tree no longer out-competing the ivy for nutrients. From this point it provides an important ecosystem service by increasing the windload on the tree and bringing it more quickly back to the earth where it can be turned back into soil to feed new life – particularly important in cool temperate climes where this process is already a relatively slow one.

† *Entropy is often viewed as an unstoppable destructive force to be feared and fought at all costs. We fear our food rotting and our cars and our houses falling apart, hence our invention of preservatives and non-biodegradable plastics. However, these things lock up vital biological nutrients, needed to make new life.*

‡ *Then again, are these bacteria separate or actually part of us? Some scientists now believe that 90% of the cells in our bodies are microbes, mainly bacteria.*

So by studying what makes nature successful we can derive ***principles of ecology*** that guide us in mimicking nature's strategies. Life is always looking to increase any beneficial interactions, and energy and resources are kept on the move.† This makes for an abundant and stable, but complex system.

Patterns in time

These are more familiar, as our lives inevitably revolve around our responses to them. Primary (driving) patterns, such as the seasons and day/night cycles, elicit secondary (responding) patterns in life by way of adaptation. In the temperate and polar regions, plants adapt to seasonal changes by growing when there is sufficient heat, light and moisture, and becoming dormant when there is a lack. This in turn drives the behaviour of the species that feed on those plants, and who have evolved three main strategies to survive through the winter months.

Each strategy takes advantage of the autumn abundance of available food. Some put on weight to stick it out foraging through the winter, some hibernate and some migrate. Perhaps our ancestors once migrated too, but for those of us now living in those regions, our permanent dwelling places offer us a new form of hibernation. Alas, we now use ancient sunlight (fossil fuels) to heat them, which may be a more energy-expensive choice than migration. Not that we have that choice anymore as our species now covers the whole Earth – there's nowhere else for us to go...

By studying temporal patterns we can now more accurately predict the arrival of spring, the best times to plant seeds for a good harvest and how much time we have to build our home before the inclement weather arrives. Sadly, many of our current human systems such as our nine to five work patterns, still take no account of seasonal and daily fluctuations in our energy levels. A redesign is in order!

† *Energy in all its forms needs to keep moving in order to avoid stagnation e.g. moving water carries more dissolved oxygen and thus supports more life.*

An Introduction to systems

A basic understanding of the way systems behave is one key to good design. Actually there's a whole inter-disciplinary theory called Systems Theory[2] dedicated to this, which studies the way complex systems behave in nature, society and science. It's an important piece of the permaculture puzzle and Howard T. Odum, an early proponent of Systems Ecology, was a clear influence on David Holmgren's early thinking.†

So when we put specific things (elements) together (into systems), how do they behave (function)? Well to consider this, let's choose an example very close to home; our own digestive system (systems are often named after their primary function).

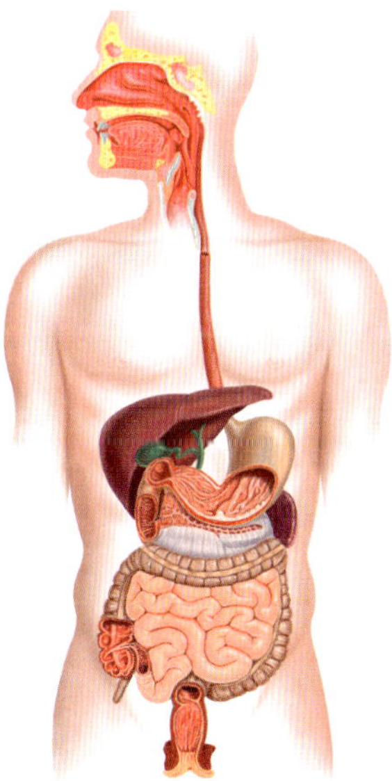

While we've all experienced how it feels when it struggles with our food choices, some of us are a little more familiar than others about the finer details of its functioning. So which elements would you say make up our digestive system? Well, it depends upon whom you ask. Have a look in some medical text books or do an image search on the Internet and you'll get a variety of answers. Some diagrams show only abdominal organs, while others include the mouth, salivary glands etc. too. This discrepancy comes from the fact that while the human body as a whole has a clearly defined edge,‡ the sub-systems (of which this is one) do not. I chose this diagram (left) as it includes the tongue, teeth and oft-forgotten nose. Our sense of smell is actually an important component of taste. This sense in turn ensures that the materials we place into our mouth are suitable for digestion, thus acting as an important filter for not just the digestive system, but the body as a whole.

† *David's book,* Permaculture – principles and pathways beyond sustainability, *is dedicated to Howard T. Odum.*

‡ *This is of course an illusion. We are in constant exchange with our environment.*

So it's actually quite difficult to define exactly what constitutes the digestive system, as we can see the edge between it and the other sub-systems of the body is rather subjective. We could make the same observation about the respiratory system, the circulatory system, the reproductive system and so on. This is because elements (or sub-systems†) within systems are often multi-functional, each performing some important, sometimes vital, functions across sub-systems and *ultimately* supporting the whole, while at the same time being supported by the whole.

Remove the digestive system from the body and it would quickly perish, as would the rest of the body left behind. So while it can be useful sometimes to conceptually sub-divide systems to make their interactions easier to consider, we mustn't lose sight of the fact that no part of a system ever exists totally in isolation. Even the smallest elements within systems could be performing functions vital to the health and stability of the overall system and also be totally dependent upon it. Imagine life without your eyes, your thumbs or even the semicircular canals in your ears that allow you to keep your balance.

So, as permaculture designers we always aim to ***make small changes***, first observing the effects that making these has on the system, and ensuring that they're beneficial, before going further. Compare this to modern corporate-driven practices that expose us all to rapid changes in our environment such as significantly raised levels of electromagnetic radiation from computers and mobile phones etc. We've not had enough time to observe the possible side-effects of fields that simply don't exist in nature. This occurs because someone wants to make a 'quick buck' before their competitors get in on the market.

Because of the complexity of systems, we often find it difficult to see the whole picture of what is going on. So another key thing we should know is that systems can sometimes behave very unpredictably (look at how unreliable weather forecasts can be, even with all the computing power now available to meteorologists).

† *While we will often talk about 'elements' within a system, those 'elements' are often systems in themselves made up of yet smaller 'elements'. Thus most 'elements' are actually sub-systems within larger systems, though on the scale we are considering, it simplifies matters to consider them as single 'elements'.*

While the elements that make up a system may all act in one direction, the combination of them all may act in a completely different way. It's also important to notice whereabouts we find elements within systems. All non-human life makes everything using ***locally sourced resources and expertise*** – and that doesn't mean food from within a 50 mile radius (try walking or even cycling that on a regular basis). No, we find species thriving only where nature provides for their needs and where their waste products can be reused. If we move species elsewhere, we risk creating an imbalance in the local ecology.

In the same way, if we move or remove any element from a system, we might throw it out of balance or stop it functioning completely. Imagine if your teeth were moved to between your stomach and intestines. We might then find our digestion performed a little less well. Things need to be in the ***right location***. Remove our teeth completely and, well some of us already know how that is. At least we can still manage to some degree without teeth, but lose our liver function and we're in real trouble...

The Yellowstone Wolves

For nearly 70 years wolves that inhabited this great nature reserve were missing – hunted and killed by humans. Over that period, elk increased in number and became sickly beasts. Young trees were failing to replace the old because of the extra grazing and the forests were dying. In 1985, 31 wolves were released back into the park and in the 25 years since elk numbers have returned to former levels. The herds are fitter because they move much more and the forests are growing back. Beavers have returned too – relying on a good growth of willow at the water's edge. In turn the dams they make have slowed river flow, reducing the loss of soil through run off and lowering the risk of downstream flooding. All of this occurred because of the wolves.

Systems also always have *functions*.† Observing a system over time allows us to determine what those functions are (and in human-designed systems, such as financial institutions and corporations, it's not always what they purport to be!).

† *In systems theory, non-human systems have 'functions' and human-designed ones have 'purposes'.*

For most systems, one of the key functions is '*to ensure its own perpetuation*'. Hence in permaculture we aim to ***obtain a yield*** and also a surplus for reinvestment. The overall functions of systems are determined not just by the functions of the elements or sub-systems that they are composed of, but also the ***interconnections*** (relationships) between them. All elements in a system can be replaced (e.g. cells in the human body, people in a university), but if the interconnections remain the same, the system will continue to function in the same way.

These interconnections are what makes any system strong (like a web), but also more complex. This makes sense of our desire to simplify things as much as possible, so we have less to think about (monoculture farming is a perfect example of this). However, simpler systems are far less resilient and more vulnerable to outside changes, like a reduction in the availability of one or more important inputs, such as oil. In contrast, a web or interconnected system can have over half its threads break and still be able to successfully harvest resources.

Feedback

Another key principle of systems is that these interconnections often create loops that feedback on themselves, which either oppose or support any change. The first kind of feedback keeps a system *in balance* and these of course abound in Nature. They keep everything vital to life, such as oxygen levels, temperature, etc. consistently at the right levels. They work by creating an opposing reaction when something changes, bringing it back to a point of balance. One example of this is how we continually adjust our posture, in order to stay upright as we walk or ride a bike, especially on a windy day. In fact when we first learn to do these things, we are simply training our body's feedback mechanisms to keep us upright.

The second kind of feedback is a *reinforcing* process, where any movement away from a point of balance stimulates a further movement away. Crossing your hands over on the handlebars of a bike demonstrates this very effectively, though speaking from personal experience, I wouldn't recommend it! An epidemic is another such scenario; the more infectious people there are, the quicker a disease spreads. After birth, a feedback mechanism is set in motion whereby we grow to the size determined by our environment to be ideal, and then at puberty the release of hormones creates feedback that keeps us at this optimum size. Balancing loops keep systems at a steady state and reinforcing ones move them from one steady state to another.

While at first, reinforcing feedback processes may not seem as common as balancing ones, we can actually find plenty of examples of them around us, many as a result of our human actions. Much of the work we have to do to repair eco-systems involves identifying and reversing destructive, reinforcing feedback loops that our relatively recent human activities have set into motion.

Spirals of erosion and degradation

Anything that we value, but that is being progressively lost; from soil to silence, from biodiversity to darkness, from trees to a sense of purpose, can be studied to help us identify the root causes of these problems.

I like to use the rather ironic example of a hole in the road, to compare permaculture with current thinking. Whereas we tend these days to think only about filling in the hole, a permaculture approach would lead us on a journey of discovery; to also find out what caused the hole to be there in the first place.

Of course, the latter approach involves extra time and effort, which is '*more than necessary*' to councils and governments who only have a few years to convince us they are worth voting for again. So each successive regime patches up the mess the last lot left behind, as cheaply as possible and with no thought for the longer-term future. Not a recipe for success.

So it's up to us as individuals to see the folly in this thinking and come up with something better. Which brings us back to evaluating why any given thing doesn't work and where we can make different choices. Take for example, using a rotavator to clear 'weeds' from an allotment plot. In the short term it does what is intended, but chops perennial 'weed' roots, propagating them in the process and also brings dormant weed seeds to the surface. This ultimately means more 'weeds', which leads to more rotavating. Until that is, the oil runs out...

Now this is a very simple reinforcing loop (which spirals more and more out of control) – 'weeds' stimulate rotavating, which in turn leads to more 'weeds', then more rotavating and so on. An obvious point of intervention here would be to choose a different method of control. However, many loops are a little more complex than this and require a bit more consideration.

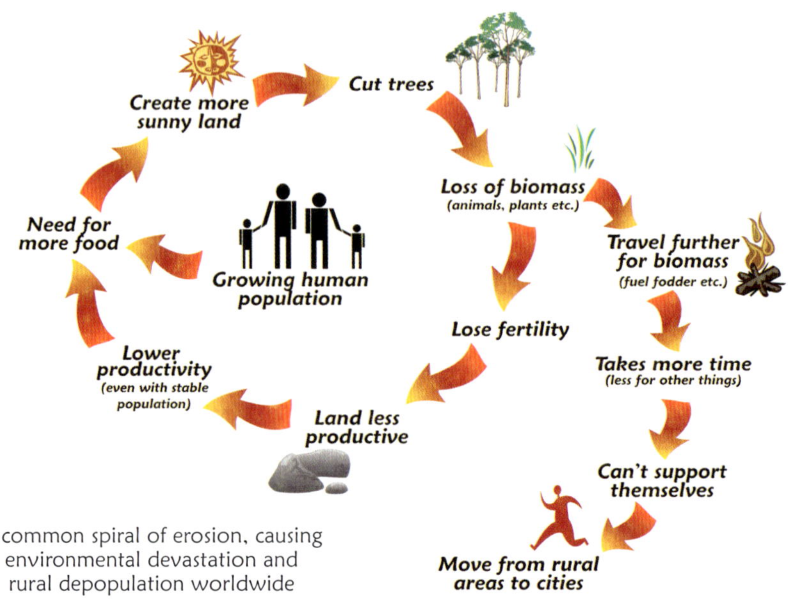

A common spiral of erosion, causing environmental devastation and rural depopulation worldwide

Start thinking about some of these loops for yourself. Pick something familiar that you see being eroded; perhaps local community, food growing knowledge, letter writing etc. and see if you can draw the spiral that has caused the problem. What interventions could you make to turn the problem around? Identifying the root causes of what we might also refer to as resource or energy leaks, provides us with one or more points at which we can begin to address them. We'll return to this later when we're figuring out what the functions of our design are going to be.

Systems game

Next time you find yourself with a group of friends (ideally at least ten of you), try this interesting exercise. You might start by standing in a circle, facing inwards. Explain that each of you needs to think of two other people in the group (but not to reveal who they are). Then explain that when the game starts they need to simply make an equilateral (equal sided) triangle with those two other people. Choose two people from the circle to demonstrate and move to form such a triangle with them both. Also point out that you could walk between them and make a triangle from the other side of them too. Of course, each of the people who are in your triangle will soon be moving to make triangles of their own!

Check that everyone is clear what they'll be doing, then start the game. After a few minutes, the group usually finds a point of stability and everyone slows to a standstill. Explain that this can represent the stability of small, fairly self-contained systems, perhaps the 'balance' each of you have been seeking relates to food supply and predator numbers in the real world. Then ask one of the group to ignore their own triangle and walk out of the cluster, but get everyone else to try to maintain theirs. It's unlikely that no one else will move (pick someone popular!). The chances are that everyone will. This demonstrates that affecting a single element within a system can throw the whole system out of balance. In the real world, changing what we might think to be fairly insignificant things (like the population of slugs or snails) can thus have unforeseen and possibly unpleasant consequences.

Principles and directives

The great value in having principles and directives is that we can use them to easily apply successful natural patterns to the many things we do. The simplest way to do this is to use the ***principles of ecology***[3] in designing gardens and farms and I'd recommend this as being a good place to start. However, we can be more creative and apply many of these same principles beyond land-based design too, something I'll give examples of in part three.

Permaculture also addresses how we think, the mess we see around us being simply a reflection of the mess in our heads. Then again, our modern lifestyle has so disconnected us from nature that perhaps we shouldn't be so surprised.† So having some ***principles of attitude*** can assist our approach and help us to see the hidden gifts in every situation. I'll be introducing some of these principles at the most relevant places in the text.

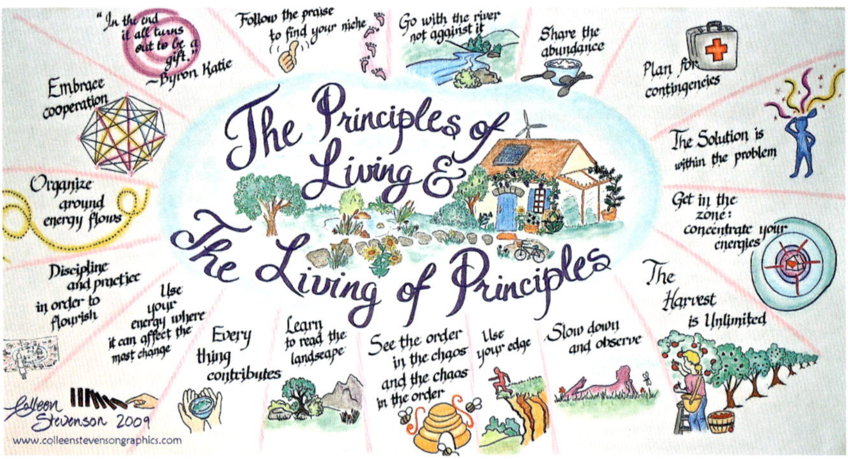

Colleen Stevenson's beautiful poster of her own version of the principles[4]

You'll find many different versions of the principles on your travels. It's not a matter of which ones are 'right', but rather which ones you're finding useful in your life.

† *In his book,* Left in the Dark, *Tony Wright makes a good case for our collective madness being simply a problem of biological engineering gone awry.*

Summary

So the key things to remember from this section are:

Observation

* *Nature is our best teacher.*
* *Great design begins with detailed observation.*

Patterns

* *...are the form of the boundary between two media.*
* *...are nature's best designs to perform certain functions.*
* *Nature seeks to optimise the edge between systems.*
* *Resources and energy like to keep on the move.*
* *Entropy increases the opportunities for life!*

Systems

* *...can be unpredictable (so make small changes).*
* *All systems have functions, directed not just by the elements within them, but also by the interconnections between them.*
* *These interconnections can create either balancing or reinforcing feedback loops.*
* *Elements can both benefit from being part of a system and also contribute vital functions to the whole.*
* *Small things can be really important!*
* *Complex systems are more resilient than simple ones.*

Spirals of erosion

* *...can be simple or complex.*
* *...are created by reinforcing feedback loops.*
* *...can be turned around at specific points of intervention.*

Principles and directives

* *...guide us in applying successful natural patterns to what we do.*
* *...positively frame the way we think about what we see.*

Effective Design

So what exactly is design and how can permaculture help us to do it well?

Design is the *conscious assembly of concepts, materials, techniques and strategies for a particular purpose.* Seeing all the exciting possibilities that permaculture offers us, it's easy to forget this and just end up throwing together a collection of 'green' technologies and techniques, only to be disappointed by the result. There's now no shortage of these 'green options' for us to choose from and we've been given the impression that as long as we behave in certain ways and buy the right products, we're doing the best we can.

But permaculture and design is about more than just choosing the right things, it's also about *how we connect them together.* Nature abounds with examples of ***beneficial relationships***, showing us the value of this strategy for long-term sustainability. So as permaculture designers, our role is to place components in the best places relative to each other, to create self-sustaining systems that also meet our needs. However, such relationships are often site-related, so we need to be able to *consciously design*; to become a 'permaculture chef', rather than simply learning to follow a recipe.

Permaculture tells us that when we design to meet our needs, we should do so in a way that supports the ecosystem as a whole, without which we are as doomed as that digestive system placed in isolation. Of course, permaculture is not a specific recipe, nor an end point. Rather it is an ongoing process of harmonious adaptation to nature's changing conditions. The design process can help us each to find and stay on our own path.

Techniques are simply *how* we do things. Choosing the right techniques depends upon a good knowledge of the limits of the eco-system in which we are working.

Strategies then add time, defining *when* we do things. This requires a knowledge of seasonal variations.

Design then adds *where* to the equation, placing things in the best location and in an optimum relationship to each other.

Everything gardens – While we may currently be making a bit of a mess of things, it's perfectly natural for us to be shaping our environment to meet our needs. Biomimicry[5] advocate Janine Benyus points out that even when our choices appear anything *but* natural, as a product of nature we can't ever do anything else. Instead she asks whether our choices are *well-adapted* ones. Permaculture gives us the tools to create systems that support not only ourselves, but future generations too and Life as a whole.

In *Permaculture – A Designer's Manual*, Bill Mollison suggests:

* *That the systems we construct should last as long as possible, and take least maintenance.*
* *These systems, fuelled by the sun, should produce not only for their own needs, but the needs of the people creating or controlling them. Thus, they are sustainable, as they sustain both themselves and those who construct them.*
* *We can use energy to construct these systems, providing that in their lifetime, they store or conserve more energy than we use to construct them or to maintain them.*

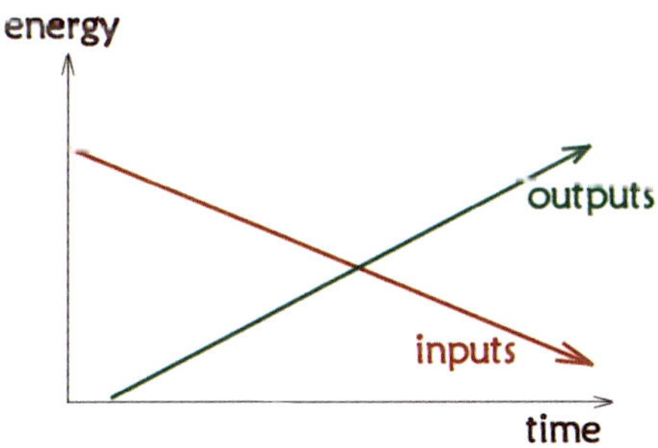

These design considerations provide us with clear criteria for how any permaculture design should perform. If we can design systems within these guidelines that meet our human needs, and at the same time support the eco-system as a whole, then we will be well on our way to a sustainable human society.

We should invest most time and energy in the establishment of a good design, so inputs decrease as time goes on. Conversely, yields may start off small but then increase steadily. At a certain point, the total energy yielded from the site exceeds the total amount invested and the system goes 'into profit'. One more thing worth remembering is that biologically we're simply an eco-system living within a larger eco-system. And that whatever we do *has consequences*.

Identifying roles within the process

Now let's consider the roles each person will take within the design process. Sometimes, as both designer *and* client, you might find yourself the only person designing and implementing a particular project. In which case, you can skip ahead a few pages to the section on ***Design Frameworks***.

Whenever you do find yourself working for or with others though, in order to avoid confusion and conflict, it helps to define everyone's roles at the beginning of the process. In such a situation, you might be:

* *Working as a sole designer or part of a team.*
* *Working for a single client or multiple clients.*
* *Also responsible for delivering the implementation of your design, either as project manager or as the workforce.*

We'll look at some tools to help with the first situation in a moment, the second we'll address in the Client Interview chapter and the third under Implementation, towards the end of part two.

Whenever designing for clients, it's advisable that you, or the agreed leader of your team, should clarify these details with them before you start, in writing in a design proposal.[6]

Such a document should:

* *Clarify the permaculture ethics and principles underlying the design process, in the context of the design.*
* *Include an overview of the design process itself.*
* *Define the agreed roles of the client(s).*
* *Outline your fees (daily rate / estimated timescale).*[†]
* *Perhaps provide short biographies of the designers.*
* *Identify the point at which your involvement ends.*

This may seem a little formal, but I've learned from experience that it's better to spend a little time clarifying these things up front, to avoid discovering a serious misunderstanding later.

† *The more often you do this, the easier it will be to estimate.*

Working as part of a team

While developing and delivering a permaculture design as part of a team can be very rewarding, it can also throw up challenges, especially for those of us not used to working with others. Here are some tools that can help smooth the process.

Inclusion

This list could be very long, so here are just a few methods I've used with small groups. Some or all of these can help ensure everyone is included and feels valued by the rest of their team.

At the beginning

Dream circle[†]

One nice way to quickly establish a good group dynamic is for each person to share how they would like the process to go for them and their best outcomes. Someone in the group records these dreams as each person shares one in turn, going around until all have been expressed. No one is interrupted, other than for clarification. At the end, all members of the group commit to doing what they can to make all these dreams come true.

Skills audit

One of the first things you should do as a team, is to identify all the skills and experience that you have between you. The best way to ensure that everyone feels valued is if they are able to contribute in their own way to the overall process. If anyone in your group is feeling that they have nothing to offer, ask them what kind of things they like to do and use this process to identify a role that they can be appreciated for.

Ongoing

Rotate roles

Ensure that throughout your design process, everyone gets to have a go if they wish at being the leader/facilitator. Likewise, also assign the roles of timekeeper, note taker and observer/vibe watcher in rotation.

† *For more details investigate John Croft's Dragon Dreaming process.*

Meeting techniques

Check-in

Start each session with a check-in to see how everyone is doing. These are also useful if the group's energy is flagging, to clarify if it's time for a break or to ensure that it's OK with everyone to carry on a little longer to finish a particular task.

Everyone speaks once before anyone speaks twice

Explains itself – ensuring that everyone has the chance to be heard at the beginning of any discussion. Ideally everyone continues to be heard all the way through the process too. Hopefully there's no need to emphasise that everyone also has the right to choose *not* to speak if they wish.

Go-round

Makes sure that everyone gets to be heard at each stage of the process. In a go-round, everyone speaks if they wish, in turn around a circle. This process needs to be managed by the group or a facilitator to ensure that people don't speak out of turn. If time is limited, you may need to define a maximum time for each contribution. You may also clarify at the start how many times you'll go around the circle. Sometimes this is just once.

Hand signals

While these are most useful for much bigger groups, having a few agreed hand signals can be useful. They allow others to indicate things like 'I agree', or 'I've a point to follow on from that' etc. without interrupting the flow of whoever is speaking.

Think and listen

If you want to give everyone the chance to consider and talk about a particular thing for a few minutes, a go-round may not always be your best option; with bigger groups it would take too much time. This technique gives each person the chance in turn to think and speak out loud without interruption. Working in pairs, one speaks while the other becomes an active listener, then after an agreed time the roles are reversed. Main points that have been identified can then be shared and recorded by the facilitator. This tool is particularly useful when you want to quickly gather ideas and opinions about anything.

Contemplation agendas

Groups often get stuck feeling that because an issue has been raised, a decision has to be made about it during the meeting. This tool allows issues to be raised for initial thoughts and set aside for contemplation and reconsideration at a later time.

Visible, open agendas

Raising points for discussion at the beginning of a meeting and writing them up where they are visible to all, allows everyone to see how well the meeting is progressing. By starting with a draft agenda and inviting additions from the group at the beginning, everyone feels involved. Having identified what needs to be covered, approximate times for each item can be assigned by agreement. With this in place, the group can decide the best order in which to address the items. If items take longer than anticipated, the group can decide how to reallocate the remaining time they have.

Parallel thinking (six hats)

Edward de Bono's concept[7] is a good way of ensuring that at any one time, everyone is coming at the process from the same direction. Very often discussions take overly long because one person is coming at it from an optimistic viewpoint, while another only sees what could go wrong. Another might simply be expressing their gut feelings about an idea while another wants to discuss all the technical information. This tool gives space for each way of looking at a proposal to be considered in turn.

These approaches are identified with six different coloured (imaginary) hats. The group first assigns a facilitator for the session, who 'wears' the *blue hat of process*. They then lead the whole group through the other 'hats' in turn, ensuring that everyone stays with the same one until all are complete with it. The other five hats can be considered in any order, with sequences suggested by de Bono for different processes. They are the *yellow hat of optimism and positivity*, the *red hat of emotion and intuition*, the *black hat of caution and criticism*, the *green hat of creativity and possibility* and the *white hat of facts and figures*. Most of us habitually think in just one or two of these ways, so it helps us to experience thinking about things from different perspectives for a change.

Conflict resolution

Hopefully you won't need to employ any of these tools!

Talking stick

The talking stick (or anything else agreed on for the purpose) is a further tool to ensure that only one person speaks at a time. It is often used to either deepen the connection between a group, or to resolve issues that may have arisen within it. Again, the group usually sits in a circle and the 'stick' is either placed in the centre of the circle for anyone to pick up, or the facilitator begins. Once started, the 'stick' may either be passed around the circle, or returned to the middle after each person speaks. Whenever someone is holding the stick, they can speak uninterrupted. Once they have finished, they pass on or return the stick to the centre. Others can only respond once they have the stick. As the process slows down interactions, it can give strong emotions time to defuse and reason the time to return before any comebacks.

'I' statements

It's common for us to make statements like 'people don't like that kind of thing', when really we mean '*I* don't like that kind of thing'. In difficult situations, it is important that everyone 'owns' their feelings about an issue. We can only truly speak for ourselves, not for others.

Non-violent communication

This is a technique developed by Marshall Rosenberg to help us communicate with greater compassion and clarity. It has much to offer and I've not nearly enough space to do it justice here. Again it emphasises the importance of owning our experience. So rather than saying something like "*you make me angry* when you do that", we might say "when you do that, I *feel* angry". The second statement separates the two things somewhat and takes back the responsibility for the experience of feeling angry. He understood that most of our responses are nothing to do with what is actually happening in the moment, but instead arise from our past experiences. There is also an emphasis upon listening with deep compassion.

Structuring the process

Now that we have a few techniques to help ensure we have a smooth running design team, let's look at some frameworks that can help us structure the design process from start to finish.

So why use a framework?

A framework lays out a pathway for us to follow, ensuring that we don't overlook anything important in the design process. There's no *one* correct permaculture framework, designers often choosing from several that have been borrowed and adapted from other disciplines. Whichever one you choose, a framework should provide you with a successful pattern that you can use to get good results more quickly than by a process of personal random discovery. Here are a few (and you may come across others), any of which might serve you well.

SADIMET

This framework has been adapted from SADI, commonly used in landscape architecture. These letters stand for the stages of ***S***urvey, ***A***nalysis, ***D***esign and ***I***mplementation; ***M***aintenance, ***E***valuation, and ***T***weaking[†] being subsequently added. Of course *the whole process* is the design, so you might consider the D to instead stand for Decisions, the point at which you commit to specific choices.

O'BREDIMET[‡]

This framework is an expanded version of BREDIM, a design framework used in industrial engineering. The letters stand for ***O***bservation, ***B***oundaries, ***R***esources, ***E***valuation, ***D***esign, ***I***mplementation, ***M***aintenance, (Re)***E***valuation, and ***T***weaking.

You can probably see the similarity between the two, the latter just adding an extra level of detail to the observation stage. Both I think provide a useful framework for designing.

† *This is of course, a technical term.*

‡ *A friend of mine once joked that I must have made this one up because when she Googled it, my websites were the only ones that were listed. However, Google just 'OBREDIM' and you'll get plenty of results. A mentor added the 'ET'.*

CEAP

Another simpler design acronym, ***C***ollect site information, ***E***valuate this information, ***A***pply permaculture principles to this information and generate a design, ***P***lan a schedule of implementation, maintenance and feedback. In my experience, those letters (especially A and P) don't really act as a good way of remembering all the information contained 'within them'.

An Australian design cycle

One I discovered on my Diploma pathway. It lays out the process nicely, but perhaps it has too many vowels to make a memorable acronym? I mean, IIEPCIORR? It just doesn't trip off the tongue like O'BREDIMET. OK, *bad example.*

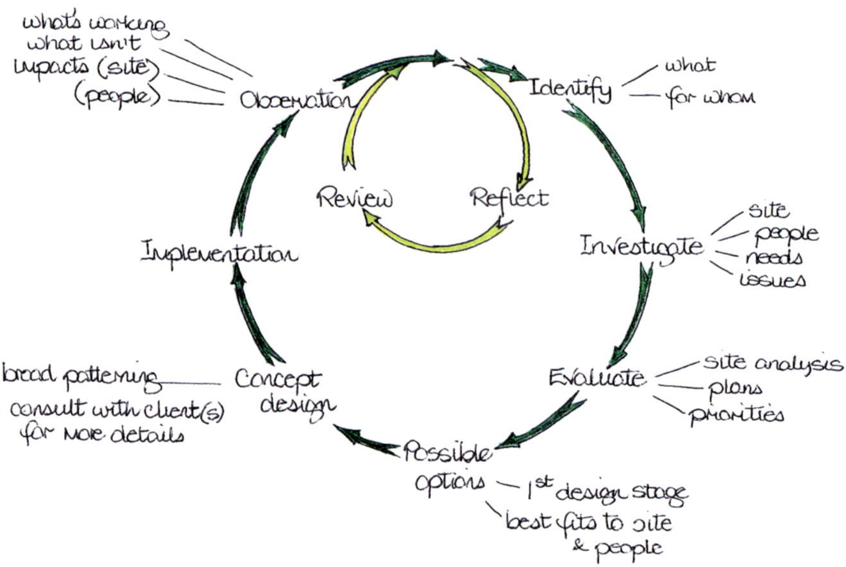

A web approach to design

Not all frameworks are linear ones. Looby Macnamara's *design web*[8] has the stages represented as 'anchor points' in a web. These can be visited in a variety of ways, making it adaptable to each situation and person. She says the process is particularly well suited to people-focused designs and we'll be looking at those in a bit more detail in part three.

The Action learning cycle

Whichever design framework we use, we should also take into account the Action learning cycle. This is literally how we learn naturally; how we learn to walk and to talk. Learning by doing.

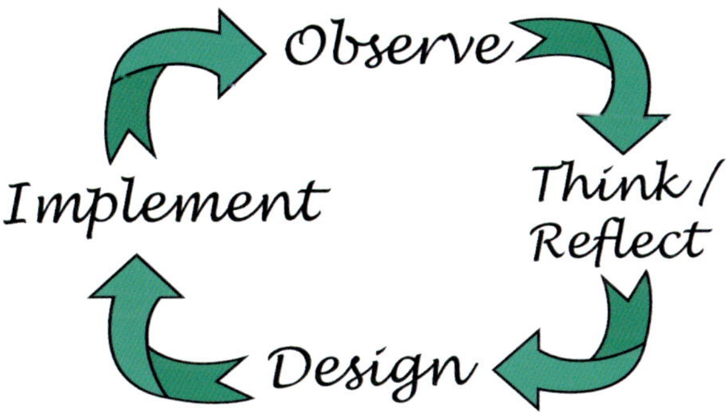

In this process we first spend time observing what's going on;

> *"How is Mum managing to stay up there?"*

Then we think about what is happening;

> *"OK, she seems to be nearly falling over, but putting a leg out each time to stop her ending up down here with me and her arms are moving the opposite way to her legs – oooeerr..."*

Next we come up with a strategy or design;

> *"Right, if I just copy what Mum's doing..."*

Finally we have to try it out to see if it works;

> *"Ooopps! I fell over again."*

Having observed how that went, we then come back to a point of reflection and the cycle starts again;

> *"So what am I still doing differently?"*

Eventually this process brings success and we learn to walk, but the mistakes we make form a vital part of our learning.

In permaculture, as long as we learn, we suffer no 'failures'.

Flowcharts

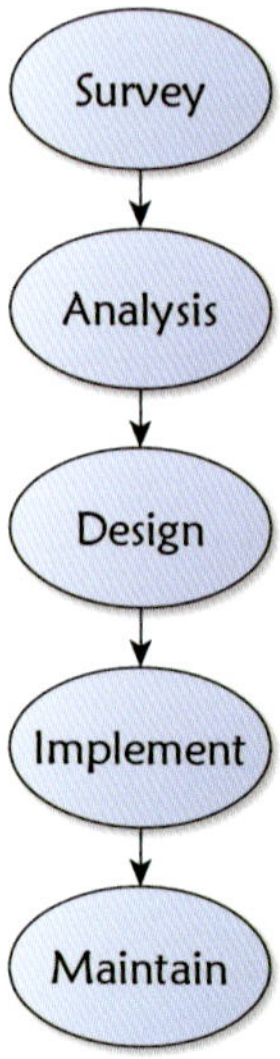

At its most basic, the design process seems like a journey from A to B, from beginning to end. In reality though, a design is never finished and continues to need attention and adjustments whether you as designer are still involved or not.

A much more accurate way of portraying any of these frameworks would be in the same circular pattern as the action learning cycle, which is no surprise, because if we look again we see that they are the same thing!

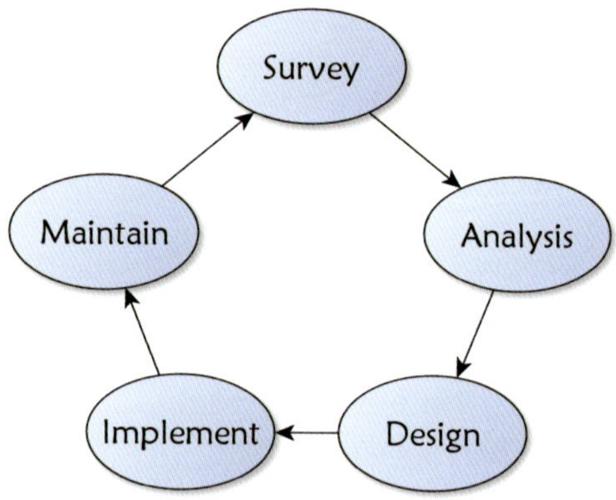

For those who like to see things spatially, each section of the design process has its own flowchart, summarising how everything in that stage flows together. They can be found at the end of each section and are also available to download¤ as part of the accompanying online resources.†

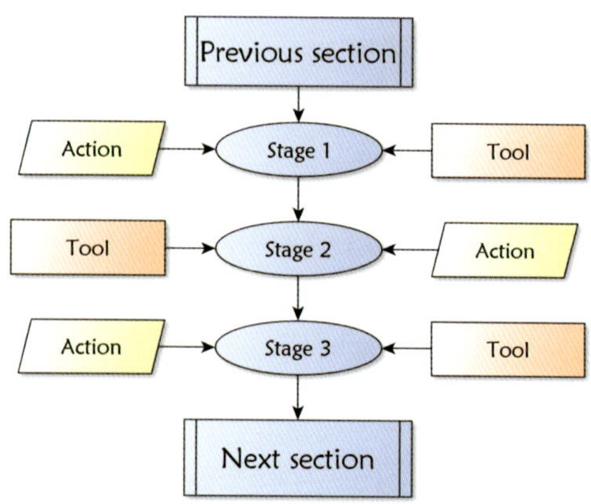

The large rectangles top and bottom represent the sections either side and the blue ellipses each stage. The yellow boxes represent the actions and orange ones the tools you might use.

† *Remember – the ¤ symbol directs you to online resources, downloadable from www.aranyagardens.co.uk/design-guide-downloads.html*

Summary

So the key things to remember about the design process are:

Defining roles

* When working for a client, clarify at the beginning of the design process, theirs and your own levels of involvement.
* Identify the point at which you hand over the responsibility for the design implementation to them.
* Put all this in writing at the start, in a design proposal.

Working as part of a design team

* Identify everyone's skills and make best use of them.
* Ensure that everyone feels heard and gets an equal chance to contribute.
* Rotate roles in meetings and discussions.
* Use the given meeting techniques to avoid needing to use those for conflict resolution.

Design frameworks

* Provide us with a successful pattern to guide our process.
* Help us to avoid missing out anything important.
* All essentially follow the broad process of:
  Survey, Analyse, Design, Implement and Maintain.
* Different frameworks might be better suited to some design processes than others.
* While they appear to be linear processes with a beginning and an end, they are effectively cyclical.
* A design is never completely finished.

The Action learning cycle

* Outlines the way in which we learn naturally.
* All the design frameworks are based upon this process.
* Any mistakes are opportunities for learning.

PART TWO

~

The Design Process

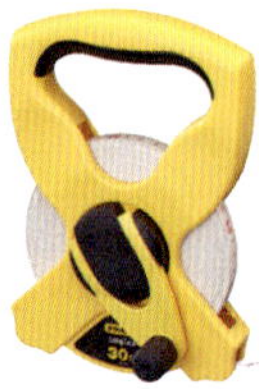

Surveying the Site

OK, so we've finally got there – the point where we can actually go out and do some designing? *Slow down, slow down*! Remember what's first? That's right, ***observation***.

"Don't just do something, stand there..."

Apologies for labouring the point, but this is by far the most common error I see being made during the main design exercise on my courses and I wouldn't want you to make it too. The first part of the process is simply to *observe* – NO designing! If you get any ideas, scribble them down on a piece of paper if you must, put them in your pocket and get back to the process of recording what you see, hear, smell, perhaps even taste and so on. How things are *right now*.

All our design frameworks begin with observing and recording and we should give good attention to this stage as it informs all the decisions we'll make later. Whereas most of us currently spend just 20% of our time planning and implementing something and then 80% maintaining it, in permaculture we set out to change those numbers around. Ironically the gift of oil has made us far more wasteful of energy, something our ancestors would never have done for long. We've used fossil fuels to replace skill with brute force in so many areas, particularly in food production. Thankfully, we haven't completely lost those skills yet and there are a few inspirational farmers who have a lot to teach us about using our resources wisely. They all understand that a well-designed system should, more or less, look after itself, though intensive food production inevitably involves a certain amount of interaction on our part, *'Harvesting as maintenance'* being the Holy Grail of design.

While this process can be applied to designing more than just landscapes (more about that in part three) we'll start off by doing this, as it's the easiest way to get a sense of the flow. In the land-based example we'll be following, the observation is made in two parts; first of the land and then of the client(s).

I say 'first the land' because personally, I like to look at a site before interviewing any client(s) in detail. This gives me an unclouded view of what I see there. The survey also often raises questions that may need further clarification, such as issues around the history of use of the site. So doing it this way around makes most sense to me. However, this isn't a hard and fast rule, and you'll have to ask the client(s) in advance about site boundaries anyway (so you know exactly *where* you're surveying!). When you do so, ask them if they already have a good map of the site that you can adapt for your own use. If they do have one it will save you mapping time later.

If you're designing for yourself, you'll already have a fairly clear idea of what you want, and if the site is familiar to you, a good sense of what is there. Then this process becomes one of clarification and expansion. It's your design process, so find the way that works best for you; be careful that you don't let your design ideas cloud your observations though! Enlisting a friend less familiar with the space, to add their own observations, can often illicit a new way of seeing familiar spaces. Such outside perspectives can help you sweep away blocks that had previously been limiting your ability to see the most obvious issues.

The main thing to remember here is that a little extra time spent observing now is going to save you a whole lot of effort later. Bill Mollison suggests observing our design site *for at least one full cycle*. For a land-based project, that means of course a *whole year*. However, if you don't have all that information to hand, there may be others who can help you fill in the gaps. Neighbours and previous owners may also know where the sunniest and shadiest areas are and where and when it floods or frosts etc. They might well have photographs that allow you to look back in time. Include these people as part of your client interview process if you can.

Maps

To design well, you always have to get onto the site, that said it helps to have a framework in which to record observations and maps can provide that for us. Hopefully your client can provide you with a map that you can use for this purpose, but if not there are still plenty of places from which you can obtain a useful one, including online services.† However, even if you're fortunate enough to be holding such a map in your hand, it's unlikely that it will be the perfect one for you just yet.

Maps are made for different reasons, so any you have will include some irrelevant things and lack other important data. What you need to know will vary of course depending upon where and what you are designing. For instance, having detailed contour information can save a lot of time surveying a large property, whereas it's unlikely you'll need to know this if you are designing a small urban garden.

Utility company map

Even having the basic outline of a site can be a helpful starting point, saving you a significant amount of surveying time in determining the boundaries, size and shape of the plot. It's important though to remember that maps are just a snapshot in time and may no longer be accurate. Only by visiting the site itself will you be able to determine this. Perhaps the only old maps that can be relied upon are those mapping geology and soils.

† *In Britain, Ordnance Survey maps are a good choice. Online services can now provide these and many other maps; I include links to some in the online resources. The most useful scale for most urban designs is around 1:1,250 to 1:2,500. Your local library may have copies of these for the site that you can view, as well as historical maps and photographs that can give clues about previous use. For rural properties, agricultural maps are often a good starting point. Land registry maps can also be helpful, though not always. Many homeowners will have these as part of their deeds. Utilities companies also often provide free maps with their supply lines mapped.*

So first identify what it is that you need to know and then investigate which maps are available that already contain this information. Remember, the right map can pay for itself many times over by saving you time spent gathering particular data. This is especially important if you're working to a budget. It's worth mentioning here that aerial photographs can also be useful, though care should be taken with those provided by free online mapping sites as they can be distorted at the point where individual photos are stitched together to create a seamless landscape. Sometimes the seam is obvious, such as when images were taken at different times of year (one issue with projecting an unnatural cloud-free landscape I suppose). You also should check such maps are up to date; at the time of writing, the Google Maps™ aerial photo of our home still shows it before it was extended by the last owner – over five years ago!

DIY mapping

There may be times when you won't have access to an existing map at a useful scale and you'll have to create your own. This is a useful skill to have anyway, so no mapmaking task is ever a waste of time. Although 100% accuracy is always worth aiming for, getting close to that can often be disproportionally time consuming. In practice 90-95% is good enough for most situations, with perhaps contouring for water management being one key exception – however nicely you ask it, water *never* flows uphill unaided!

Urban garden designs usually include a considerable amount of detail, planting schemes etc. and so need to be surveyed accurately enough to ensure that any proposed garden beds and paths will all fit into the given space. Broader scale (e.g. farm) designs are often more pattern based, addressing how separate systems can be connected together in the most effective way. Slight inaccuracies in DIY mapping may only lead to the need to plant a few more trees in a hedge line, and it's worth remembering that the final canopy sizes of trees given in books will always be approximate and site dependent. That said, paying for an accurately contoured map can allow you to for instance, design and precisely lay out a Keyline® system† on the ground using an affordable GPS unit.

† *A water harvesting and soil building strategy developed by P. A. Yeomans.*

Making base and field maps

As well as needing a map to communicate our design ideas, we also need one onto which we can record the information we'll be gathering. We call this our *base map*. It should be simple, mapping site boundaries and 'fixed' elements like buildings, roads, significant bodies of water and large trees. In addition it should include a scale for the map, the direction of north, the place and the date.

Create a base map from an existing map

If you've already found a good map, this should be a relatively simple process, but you'll still have to pick out the information you need and probably enlarge it too. Assuming the map is still accurate (and they won't always be), you should first scale it up to a useful size. You're going to need both a base map that you can present your ideas on later and a smaller field map that you can record information onto whilst on site. These will almost certainly need to be at different scales. Site work will usually involve using a clipboard (commonly A4, though A3 is also available), whereas design presentations, especially to a group of people, are better done at a larger scale (A3 or above).

Extend the boundaries

A good method for expanding or simply copying a map is to use a large window as a makeshift light box. Obviously this only works during the day! Tape your existing (smaller) map onto the inside of the window and then your larger sheet of paper fairly centrally over the top, so you can see through the big sheet to the small map underneath. Place a dot on your big paper somewhere over the middle of the smaller map beneath; this will act as the reference point for all your measurements in expanding the map. First though we need to determine how much we can scale up and that will depend upon the relative sizes of our paper.

We do this by measuring the distance between your central dot and a point on the boundary of the small map. Then decide how many times you can multiply that distance and still have the expanded version of the map fit on the bigger sheet. In the example opposite we are doubling our measurement.

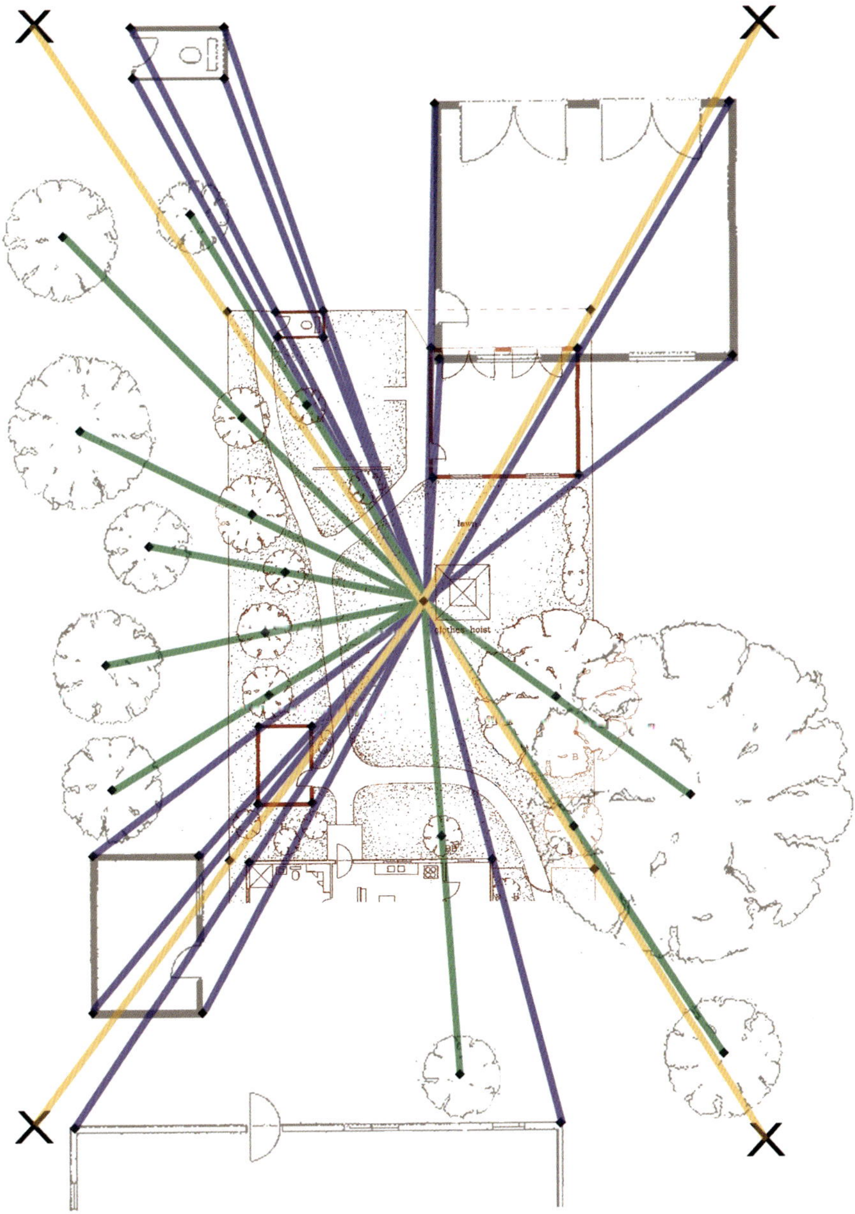

Expanding a map from a central point, in this case doubling its scale

So for instance, if you were expanding your map from A4 to A2, you'd see that you can extend a line from the central dot, through a boundary point on the small map and on to a point twice that distance, still fitting the longer line on the bigger sheet. In our example, if it were 15cm from the central dot to one point on the small map boundary, you could also fit in another 15cm beyond that on the bigger sheet (30cm in total). In order to scale up the map you would simply multiply all your other distances from the centre to the boundary points by two. Remember though to leave room on the expanded map for a key and scale etc., so don't go right to the edge of the paper all the way around. Once you've extended each boundary point (and those of any other fixed elements like buildings), join them together. If you've something resembling a large version of what you started with, well done! If not, you did something a little different than I'd hoped and learned one way not to do it. Have another go.

Don't forget to enlarge the scale. One important thing to know is that if your original map scale is written as a ratio (such as 1:100), then it will be wrong as soon as you enlarge or shrink it. For this reason, expressing scale in a 'ruler' form (see example below) which expands with the map, is always the safest option.

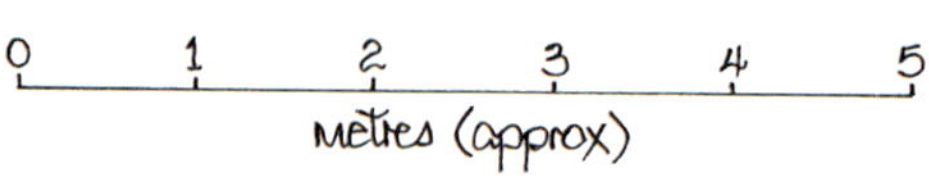

You'll discover this method works best for mapping human-designed systems where there are lots of corners and straight edges. Trees can be fairly easily mapped too as they have a centre, but more organic curves like the course of a stream or the edge of a pond will need many more points to be plotted.

Enlarge the grid

This next technique can help fill in those gaps fairly quickly. Any map with grid lines can be enlarged by hand; take your larger piece of paper and draw an expanded version of the grid on it. Then copy the information you need, one grid box at a time, into the larger grid as shown below.

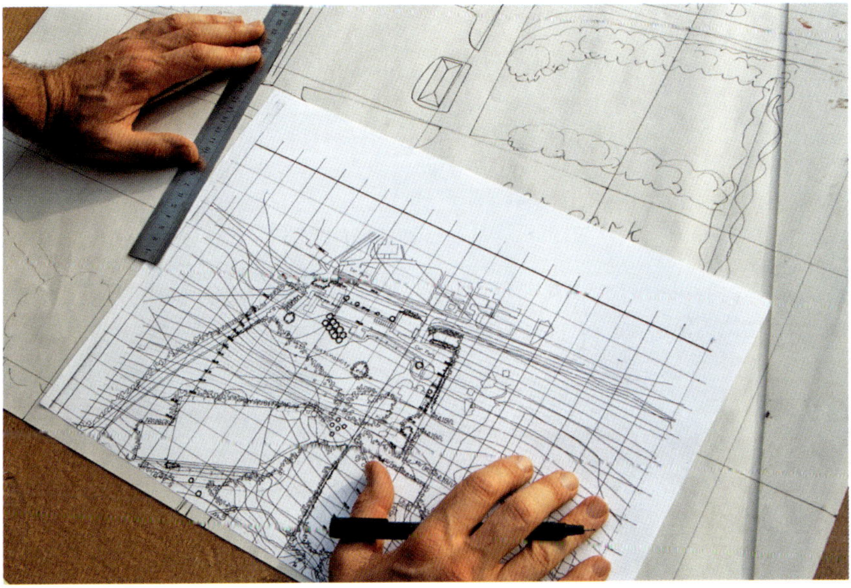

Visit your local copy shop

Sometimes modern technology can be really helpful and worth using if the end goal justifies it. Just don't create systems that rely upon it! For instance, photocopiers are very good at taking a map at one scale and quickly enlarging it to another (though travelling to a copy shop and back may take time and energy). To use this method, identify the relevant area on your original map and have it enlarged to fill a whole page.† You can then use this as the basis of a new map if the original contains a whole host of superfluous information. Simply trace over it onto another sheet of paper (the window technique works well for this too) or use the gridding process described above. Don't forget to add a scale and a north arrow if you've copied just one area from a bigger map and leave room for a key too.

† *This could be to just A4 and then enlarged again later, or straight to your final presentation scale and later reduced again for field mapping use.*

A base map from your own measurements

If you have to create a map completely from scratch, you've a bit more work to do. With a moderate investment and the right know-how, GPS can make plotting points onto a map fairly easy, but I haven't the space to cover how to do that here. For this process I'm assuming that your measuring tools are a little more rudimentary (and perhaps more appropriate). Simpler technologies tend to be more reliable and/or easier to replace if they fail. Our own limbs are a given length and we can use them to make remarkably accurate measurements. Of course everyone is different, but once we're familiar with our own stride length over different terrains and gradients, we've a measure always with us that gives us 90-95% accuracy. That may not sound so good by laser standards but it's fine when designing for plants and trees. A Bunyip† water level may not be an exciting technology, but its reliability comes from an unchanging law of physics – that water always finds its own level.

Sketch out a field map

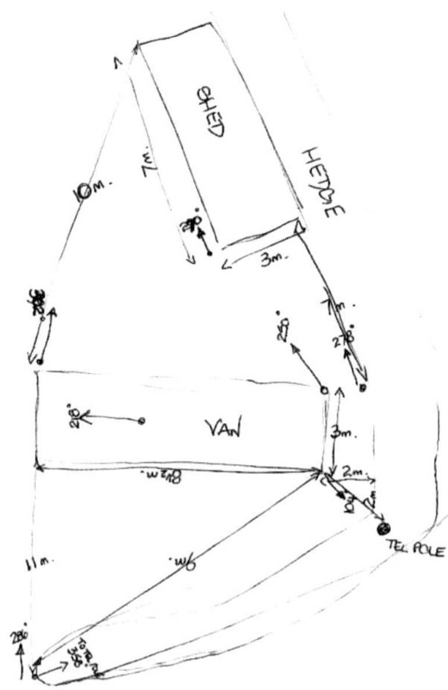

To start with, sketch out a rough field map, without worrying too much about accuracy. You'll use it shortly to record distances or bearings between everything, giving you the data you'll need later to make a more accurate map. Remember that where you stand will alter your perspective, making the site look bigger side to side than front to back. Only from above can you get a true picture. So stand in the middle and make a simple pencil sketch, or if this is difficult choose the mid point of one boundary and then adjust your sketch based on a second viewpoint at approximately right angles to the first. This should give you a reasonably good starting point.

† *Named after a fierce stick-limbed aboriginal dreamtime creature. The water level comprises a clear hose attached to two tall measuring sticks.*

Choose your baseline

Next, plot the key fixed points on the site such as buildings, gateways, fencing corner posts, telegraph poles or big trees. Start by choosing two points, perhaps along one side and a good distance apart, from which you can measure everything else. If need be, drive in two posts yourself for this purpose. For most urban garden designs you might choose two corners of an adjacent building. Such walls are often straight, making it easy to measure between those points and so providing you with a useful *baseline* for your mapping.

Measure distances or take bearings

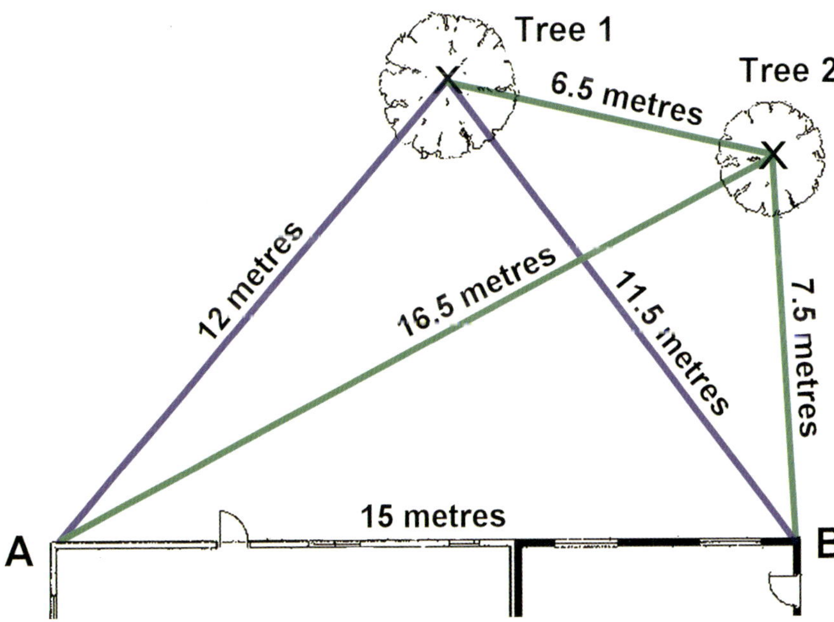

In this diagram we see a building being used as a baseline for the survey, corners A and B being measured as 15m apart. We know such a building will be easy to draw using one or more straight lines, so it will be a good way to start our map. From here, the simplest method of pinpointing the other elements on the site, such as Trees 1 and 2, is to measure their distances from each corner, A and B. You can use a site tape or pacing, see the online resources for a simple pace conversion table.¤

This will later allow us to use trilateration† to map all their positions accurately. This technique is most accurate where the directions of the measurements to any element are closest to being at right angles to each other, like to Tree 1 in the previous example. Therefore it's useful to choose another point that can be fairly accurately located from A and B (such as Tree 1), that can act as a third reference point. Hence in addition to measuring Tree 2 from A and B (measurements that are at quite an acute angle to each other), we can also measure from Tree 1.

As you do this, you'll notice that your third reference point is most useful if it's close to the other side of the site. So in order to measure its position accurately, we're better off starting with a baseline along the *longer* dimension of the site. The down side of this is that any elements close to the middle of this baseline (like our shrub below) will have measurements crossing at very obtuse angles, the opposite problem that we had in locating Tree 2, but equally unhelpful.

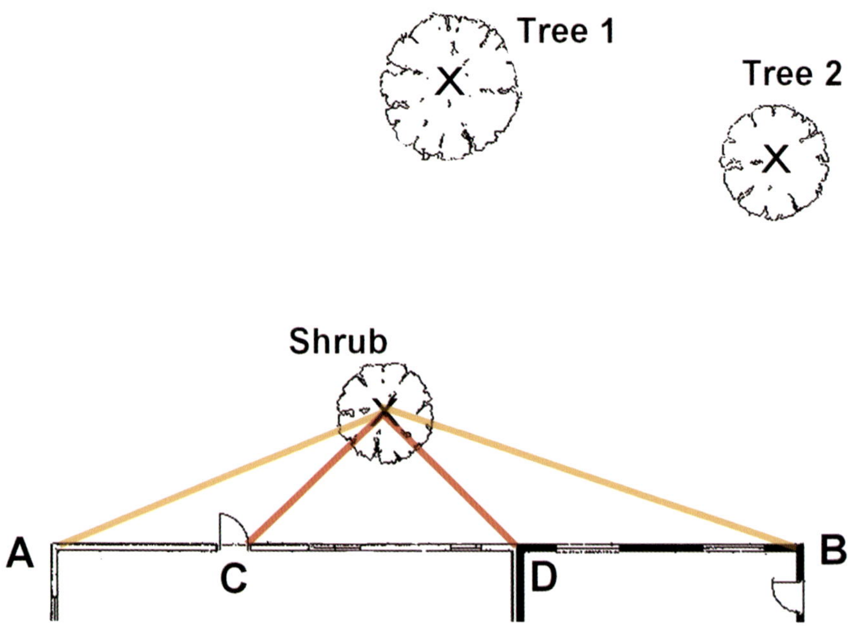

† *The determination of absolute or relative locations of points by measurement of distances, using the geometry (in 2 dimensions) of triangles.*

In this case the measurements from A and B cross at an obtuse angle, reducing again the accuracy of locating it. So our ideal baseline might have other points along its length that we can use for pinpointing closer elements. In the case of our building, the edges of doors and windows would be ideal. In our example we've also measured to the shrub from C, the side of the door and D, the join between the house and the garage. This gives us close to a right angle again. Of course more measuring takes more time, but personally, I'm inclined to invest that time while surveying a site. Each measurement I make provides me with a further check when I'm later creating the base map.

In most circumstances making distance measurements should be enough to enable you to make a fairly accurate base map later. However, there are times when it's not possible to easily measure between two points, such as when there's a large pond in between. This is when it's useful to be able to use a sighting compass. As long as you have a sight line, you can always take a bearing. A compass can, amongst other things, save you a lot of time walking back and forth across big sites.

So rather than measuring the distance between elements, we could instead take bearings for later triangulation.[†] We can do this because the Earth provides us with a reliable reference point that we can use: *magnetic north*. At least it's reliable across our site – the position of the magnetic north pole slowly shifts about and is different from the true pole,[‡] the angle between them (called magnetic declination) varying across the globe.

So starting again from one end of our baseline, say A, we take a bearing to any element we wish to map, e.g. Tree 1 in the next diagram. Without moving we can then take bearings to all the other visible elements too. Only once we have finished (and taken photos – more about that shortly) will we need to walk to the other end of our baseline and repeat the process. This can make surveying a much quicker process. However the same issue applies about placements being more accurate where bearings cross at close to right angles.

† *The process of determining the location of a point by measuring angles to it from known points at either end of a fixed baseline.*

‡ *The direction along the earth's surface towards the geographic north pole.*

It's also worth mentioning though that the process of walking about a lot across a site increases the likelihood of you noticing any number of important things you might otherwise miss.

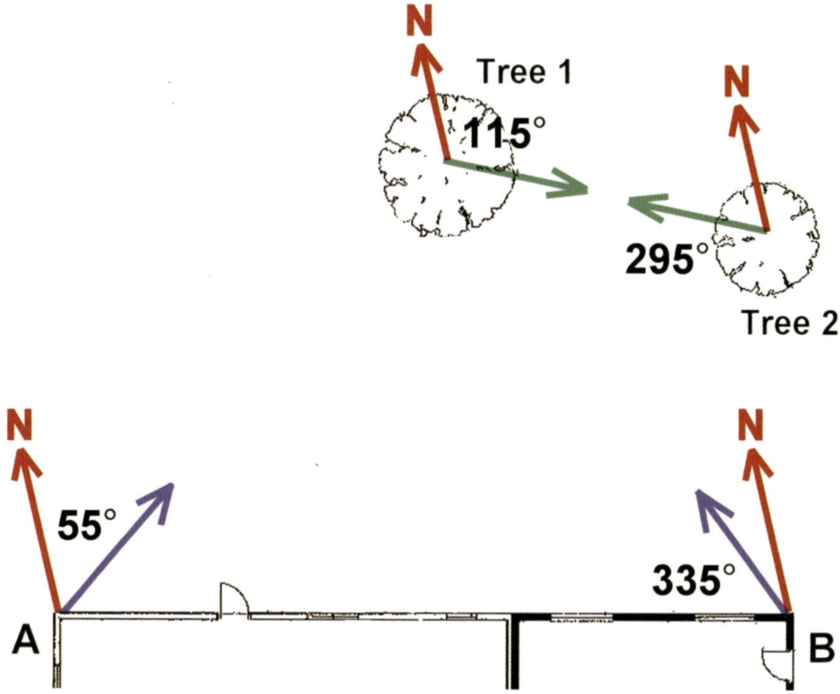

So it's useful once more to have a third reference point, ideally on a different boundary, and already pinpointed from our baseline. In this case we'll assume that Tree 2 is along a straight fence line at right angles to the building and easy to pinpoint. Adding a third bearing from there, enables us to pinpoint Tree 1 even more accurately.

Note that the bearing you take will be 180° different from that taken from the opposite viewpoint (i.e. from Tree 1 to Tree 2), so mark its direction on your field map with an arrow, as shown in the diagram. There's one more thing you should be aware of that applies to both measuring distances and taking bearings; if you now try to locate another element using Trees 1 and 2 as a baseline, you'll multiply any errors in their positioning. So try to measure everything in relation to just a few, easily located reference points.

I'm not going to explain how to use a sighting compass here, as that would take up too much space and if you buy one it will come with instructions. All I'm going to say is that as long as you're careful not to hold the compass too close to metal objects (including those you are carrying such as phones, PDAs, torches and clipboard clips), taking bearings in this manner is a quick way of pinpointing elements across a site.

Mapping non-point elements

I mentioned earlier that when expanding a map, any non-point element (such as a stream or pond) is more difficult to plot. The same applies when mapping a site from scratch. So far we've been measuring elements that we can pinpoint, whereas the curved bank of a pond is less easy to define. We might be able to map the point at which another element, such as a stream or a jetty interacts with it, but what about the rest of its edge? Assuming you don't have a GPS unit (this is where technology can really save you time), or an aerial photo that can give you a sense of the outline of, say a large pond or small lake, we need to consider some lower-tech methods.

We can start by pacing the circumference of a lake. We can measure the shortest distance between our already mapped points and its edge. We can take bearings from those points towards the left and right edges of the lake and to points like a jetty. We can also spend a little time sketching out its shape, from different angles. If you've someone else to work with, and your measuring tape is long enough, you could walk either side of the lake, measuring across it in different places.

These measurements combined should be enough to help you plot its shape on your map later. For large bodies of water, aerial photos can clarify the shape, though it's better to take plenty of measurements while on site in case you can't get hold of one that's sufficiently up to date.

Recording contours and any significant slope

If the site is small and essentially flat, it may not be necessary to give this much attention. However, if there are neighbouring slopes channelling water and materials towards or away from the site, these certainly need to be recorded and accounted for. Larger sites and small sloping sites will certainly need to have any gradients surveyed and contours marked.

If so:

- *Use a level (e.g. A-frame, Bunyip water level, laser or dumpy level) to identify important contours and any difference in height between key elements on the site, such as the fall of streams. Remember a body of water like a lake gives you a handy level reference around its edge.*
- *Mark this information onto a new copy of your base map or an overlay.† Ideally also sketch out a cross-section of the site through different areas of planting and terrain and mark its line upon your base map.*

Marking out a contour using an A frame

I always choose a water level over an A-frame, because of its speed of use. However, the parts for a makeshift A-frame are more easily found in nature; three branches, some vine and a rock suffice if nothing else is available. While more costly, an outdoor rotating laser level allows you to mark contours quickly and over longer distances, GPS can then be used to record these coordinates for later transfer into computer-based mapping programs. While obtaining a good map in advance, can save you a lot of contour surveying on larger sites, if your design calls for Keyline® ploughing or swales you'll be marking these lines out on the ground later anyway.

† *Semi-transparent sheets, such as tracing paper containing additional information.*

Take plenty of photos (or video)

While I measure as many distances and bearings as I can while on site, I've sometimes found when creating my base map later, that I've missed a vital measurement. So if the site is any significant distance from my home, I'll take plenty of photos across it, from many different angles, especially from the fixed points I've chosen (making notes where each photo is taken from). These help me later with any uncertainties in my map-making, showing me valuable sight lines and saving me the need for a return visit to clarify anything I forgot to record the first time around. More about that later. Of course, if your camera (or phone) has video capabilities too you can record some panoramas at the same time.

Useful site surveying tools

(for mapping and gathering site information)

Simple tools (that many of us own or can make or borrow):

- Clipboard (A4 or A3)
- Clear plastic sheet/bag to keep paper dry
- Plenty of paper, or copies of the base map
- Tracing paper for information overlays
- Pens, pencils, sharpener, eraser
- Measuring tapes, rope with knots, or pacing chart
- Magnetic compass (sighting versions are best)
- Water level (small syringe useful for calibrating) or A-frame
- Soil testing charts (see online resources)
- Spades (two makes soil sampling easier)
- Jars for soil samples (tall are better, with good seals on lids)
- Bags for collecting samples
- Camera (digital is best) and spare batteries

More specialised tools:

- pH/salinity testing kits
- Sun compass
- Dumpy or outdoor rotating laser level kit
- Hand held GPS unit
- Video camera (for recording dynamic events such as strong winds or heavy rainfall)

Drawing your base map

If you're close enough to do so, take your field map home, create an accurate base map and then return to the site to record other site information. However, if you're having to create your field map *and* do the whole survey in just one visit, try to make your map as accurate as you can and use plenty of tracing paper for overlays. If this is the case, skip ahead now to 'Recording site information' (p56) and return here afterwards.

Decide your map scale

Taking our field map, we first need to determine the *scale* of our more accurate base map. This we decide by identifying the longest dimension of the site and the size of the paper we are intending to use for our drawing. In my mobile home design – the rough field map example (p38), the approximate site dimensions were 25m by 12m. As the longest dimension was significantly bigger than the shorter one, I used this to guide my choice of scale. On that occasion I decided to draw my base map onto A3 paper (420mm x 297mm). As my design was eventually to be scanned and displayed online, A3 seemed big enough. If you need it to be bigger to show to a group of people and you won't have access to a projector, A1 (flipchart size) paper would be more appropriate.

Useful mapmaking tools

- Drawing board
- Paper, tracing paper
- Pens (pref. technical drawing), pencils, including colours
- Eraser, sharpener
- Pair of compasses (for drawing arcs)
- Protractor, set square
- Ruler (ideally one with a selection of scales on it)
- Set of stencils
- Or for the technologically minded, Computer Aided Design combined with data points from a hand held GPS unit.

Normally I might have divided the longest measurement (25m) into the longest side of the paper (420mm), but I noticed that I could fit the longest dimension onto the page *diagonally*, if I chose a scale of 1:50 (25m into 500mm). Interestingly, this also allowed my north arrow to point to the top of the page, always a good convention to follow to avoid confusion. Then having decided upon a scale, I divided the real life distances by the scale ratio (in this case 50), to calculate the measurements on my drawing.

It's very easy to forget to scale down one or more measurements if you do them one at a time. In my case, forgetting to divide 10m by the scale of 50 might leave me plotting a 10cm line instead of the 20cm it should be. So I recommend that you take your field map and do all the scale calculations first, writing them in a different coloured pen to distinguish them. Alternatively, a scale ruler allows you instead to plot everything using your original site measurements.

Draw in your chosen baseline

Next we need to create a reference from which all our other measurements can be plotted. Having calculated the scaled-down length of your baseline and the best place for it on your map, draw it in. Ideally, at this point you'd align your paper to true north first (this is convention), even though most of your measurements will be plotted from your baseline.

Plot the fixed elements

So, marking lightly at first with a pencil until you're sure your markings are correct, start adding details to your map. The methods are different, depending upon whether you took bearings or measured distances. If you took bearings, you're going to need a protractor. If instead you surveyed the site by taking measurements from either end of your baseline to each element, you'll need a pair of compasses for drawing arcs. If you don't have one, you can achieve similar results by using a long strip of paper. I'll assume you have compasses.

From distances: First identify the scaled-down distance of your first element (let's say Tree 1) from corner A. Open the compasses to that length, anchor the point at A on your map and draw an arc across the approximate location of Tree 1.

Now do the same, only this time for the scaled-down measurement from corner B. If you have done this correctly, the two arcs should cross as in the diagram below.

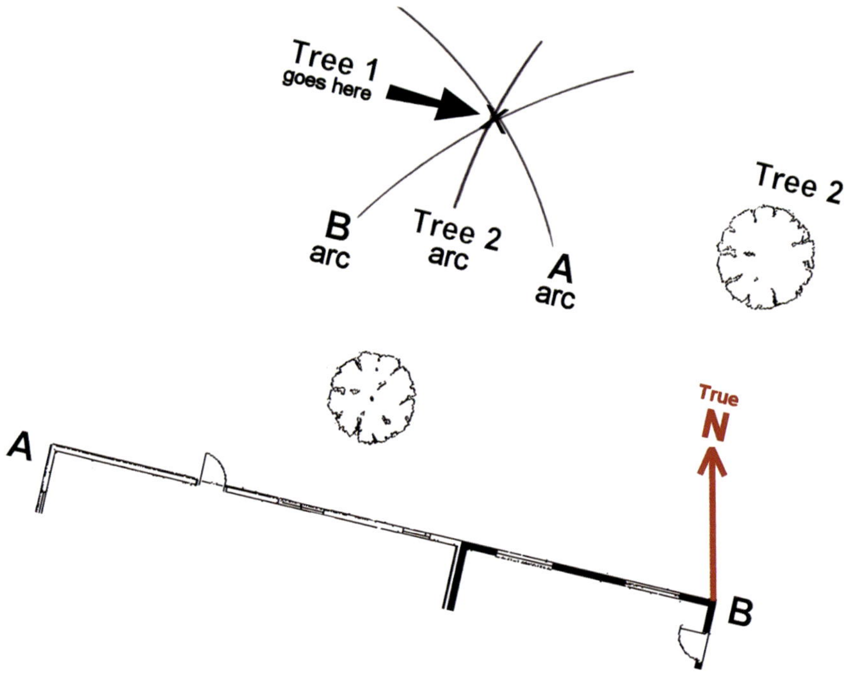

Once more, the degree of accuracy is greater if these arcs cross at close to a right angle. If this is not the case we can draw another arc in reference to a third known point (Tree 2 in our diagram). However, it would be highly unlikely for these three arcs to cross in exactly the same place. More than likely they will form a triangle. The actual position is probably somewhere in the middle of this. Repeat this same process for all the other site elements you have measured.

Try to limit your reference points to the ends of your baseline and a third chosen reference if you can. Each new reference point, especially if not being accurately pinpointed from both ends of your baseline, increases the risk of compounded errors.

Hopefully at the end of your plotting process, everything will look proportional. If so, it's time for the next stage.

From bearings: If instead you surveyed by taking bearings, you'll need a protractor for plotting out your points. One key advantage of this method is that once you have scaled-down your baseline, there are no more scaling calculations to do.

Instead the risk of errors comes from drawing bearings in relation to the wrong reference direction. To minimise the risk of error, align your map with north to the top. I'm talking now about *true* north, not magnetic north (which is where a compass points).

Magnetic north moves about over time and its direction also varies depending upon whereabouts you are. In some parts of the world the difference (which is called magnetic declination) can be quite significant.

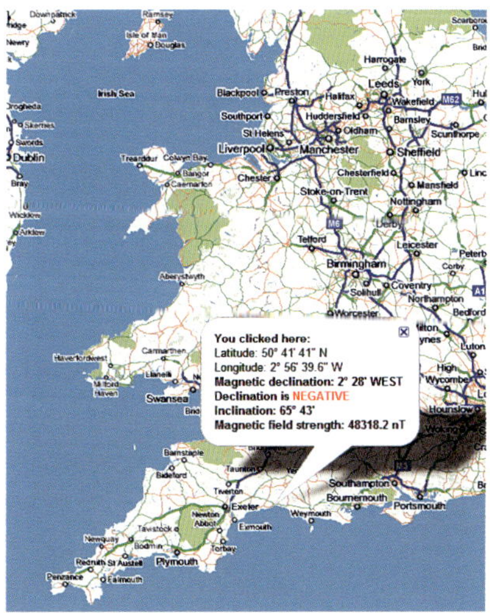

Thankfully, the good old World Wide Web tells you the current magnetic declination for any location.† You just click on a world map and a pop up box tells you the number you need to know. Once you've got this figure, what do you do with it? Well, to turn your magnetic north bearings into true north bearings before plotting them, you'll need to *add* an easterly (positive) declination or *subtract* a westerly (negative) one. So by way of an example, the declination where I'm living now is about 2½° west. With a small declination like this you might just ignore it to simplify matters; if you were somewhere it was 10°, 15°, or even as much as 40°, you'd want to adjust for it.

So in this case where I live, the true bearing from Tree 2 to Tree 1 in the diagram overleaf would be:

295° minus (as it's westerly) 2½° = 292½°

† *The online resources also include links to websites that have such information.*

If you have to adjust for magnetic declination, don't forget to do this sum for *every* bearing you take, or you'll end up plotting things in the wrong places. As before I would write the adjusted bearings in a different colour to limit the risk of this happening.

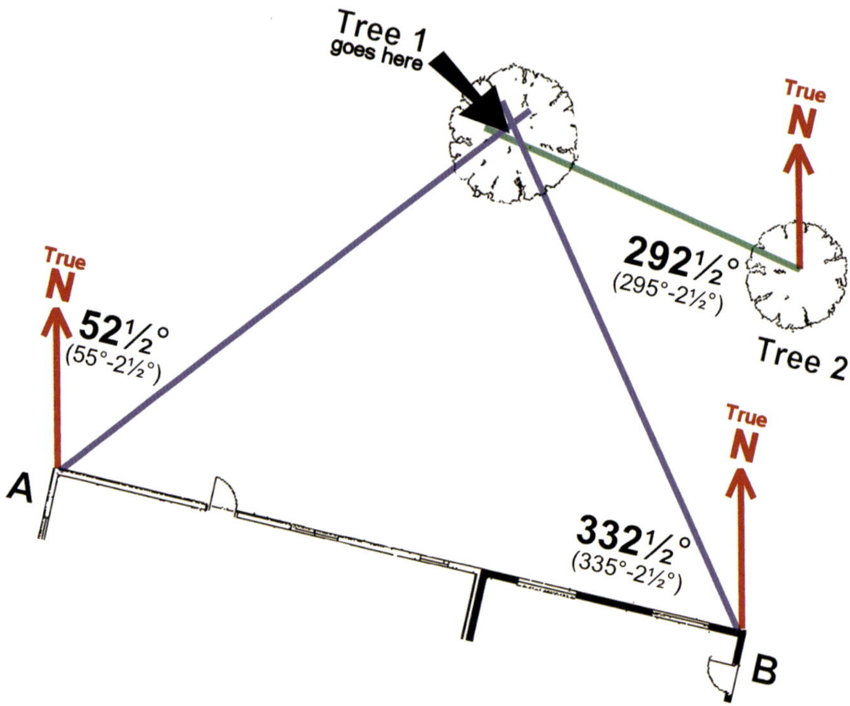

So decide whether a portrait or landscape format suits the site alignment best and use the edges of your paper as the reference for plotting your bearings. Perhaps you can now see another reason why grids on maps are popular? Having aligned true north to the top of the page and adjusted your bearings (as we've done in our diagram), all you need to do is to use the protractor to draw in the bearing directions to each element from points A and B. Calculate each bearing by measuring clockwise around a circle from north, which is at 0°. Mark each element's location where the pairs of lines cross. As before, if you want to be extra sure, you can use a third bearing to create a triangle, inside which you can plot a point.

Plotting non-point elements

We can use a combination of these two techniques to plot the lake we were mapping earlier. Effectively we are just plotting a series of points that we'll join up to make our shape. If you had the skill to take bearings, plot these on your map first. The diagram below shows how two bearings from each point (the coloured lines) quickly define the space inside which the lake must lie.

The additional distance measurement that we took from Tree 1 helps us to define how far inside that space the north bank lies and we can do the same with the other measurements we took from our other mapped points. Having sketched an outline on site, we can now transfer that inside this space.

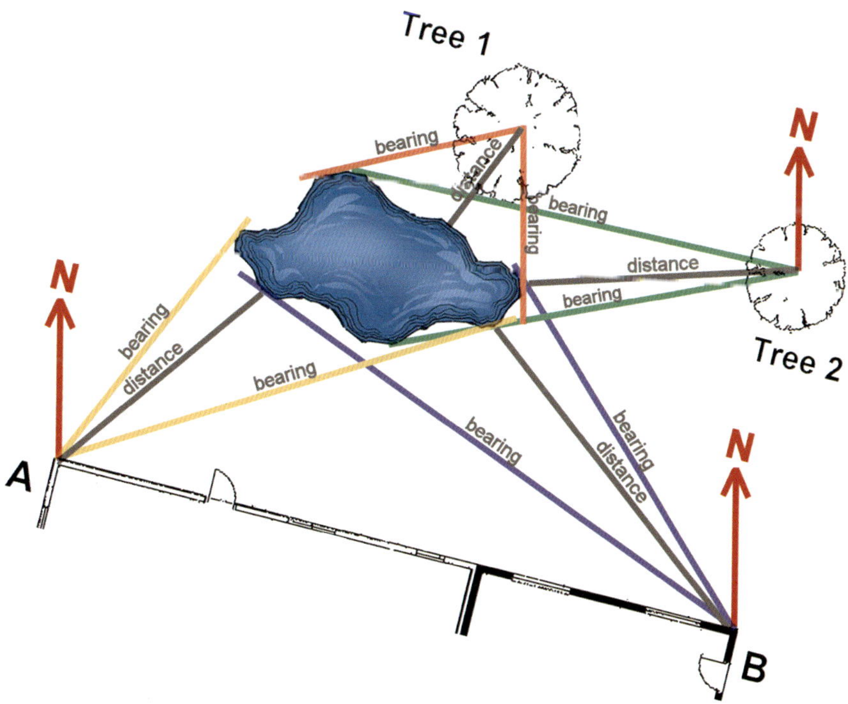

As an extra, you can scale down the circumference that you measured, cut a piece of thread to that length, join its ends together and lay it out on your map until the shape looks right before tracing the outline onto the paper.

What if I forgot to measure something?

Don't worry; making mistakes is the quickest way to learn. This is where our photos help (you did take some didn't you?). In our diagram below, photos taken across the site from our three reference points A, B and Tree 2, can help us pinpoint Tree 3 if we forgot to take measurements or bearings for it.

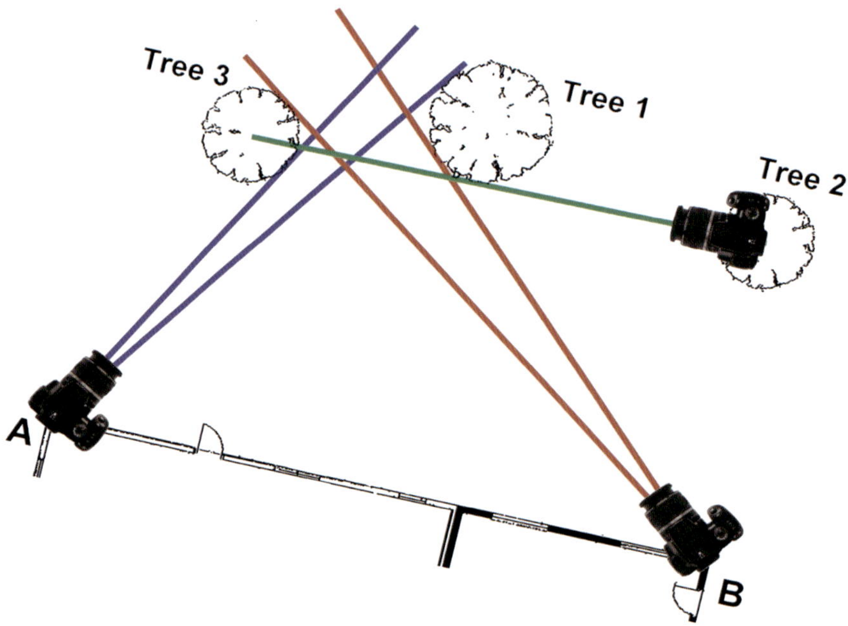

The photo from point A shows us that Tree 3 is to the left of Tree 1, the gap between them along that line of sight being about half the width of Tree 3's canopy. From B, Tree 3 is again seen to the left of Tree 1, but this time the gap is more like the full width of Tree 3's canopy. The photo from Tree 2 shows us that the trunk of Tree 3 aligns with the outside of Tree 1's canopy. Therefore, even without measurements, these photos still help us plot Tree 3 fairly accurately onto our map. Try it!

The same technique can also help us to plot non-point objects like a pond if were unable to take bearings. We can again use sight lines from known points, past its different edges, allowing us to plot points on its boundary in relation to other elements.

Join the dots, finish the base map

If after having plotted all your points out, the distances and angles look correct, join the points together to create your map. It should at this point look correctly proportioned – if not, recheck your measurements, it can be easy to miscalculate when scaling down distances. Once you are happy with your map in pencil, draw over it with a good pen.

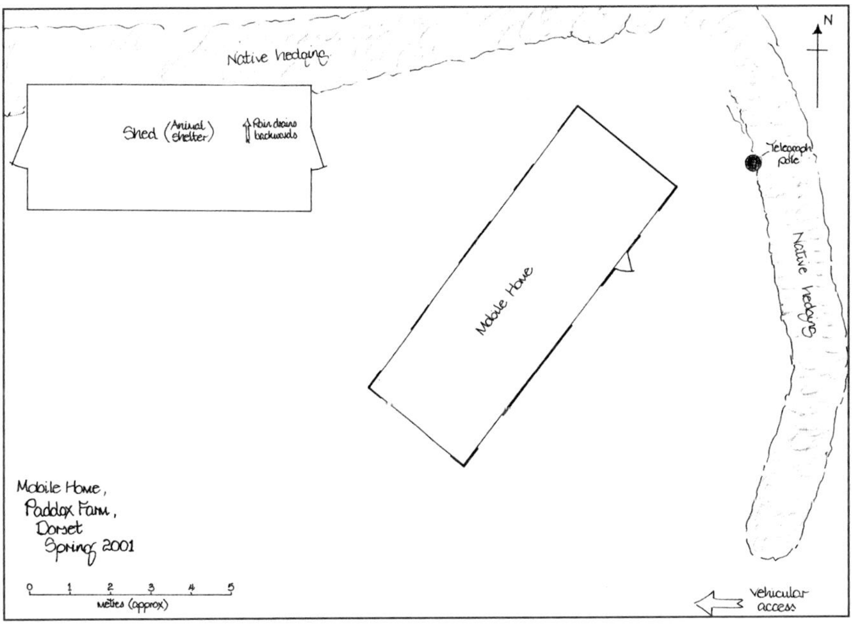

Here's the final base map (above) for my mobile home design. It includes only the 'unmovable' elements, the rest of the space being under consideration. All the other site information was plotted onto overlays. Compare it with my scruffy looking field map (p38). As you can see, you don't need an accurate field sketch, as long as you record enough measurements and/or bearings. This design was for my home garden, so photos were less important that time. Remember their great value though when surveying away from home. The cost of travelling back to a site, both in time and money, can be substantial. Prepare well for your survey trip and get what you need first time. Now make *at least one copy* of your base map in case you accidentally ruin the original.

Summary

The key things to remember about creating a base map are:

Surveying the site

* *Observe, but don't start designing yet!*
* *You may only see a site in a particular season. Clients and neighbours may be able fill in the gaps though, so make a note of any questions you have for them.*
* *A friend can offer a different perspective on your own site.*
* *Existing maps can save you a lot of surveying time, so it's even worth paying for one if need be. Aerial photos can be useful too. Be aware that either could be out of date.*

Creating base and field maps

* *Scale up existing maps wherever you can to save time.*
* *Making your own field map is time spent learning a useful skill. Measure distances or take bearings from a baseline.*
* *The degree of accuracy required depends on what you plan to do. 90-95% correct is usually good enough.*
* *Survey slopes where relevant.*
* *The simplest tools are usually the most reliable, though more technological solutions can save you time.*
* *Take photos across the site from key points.*

Drawing your base map

* *A base map shows just the main fixed elements on a site. You can use overlays to record everything else.*
* *Decide on your scale, dividing your longest edge into the long dimension of your paper, but leaving room for a key and other information on your sheet.*
* *Align your map to true north.*
* *Start by marking out the baseline and plot all other points from there. When you're happy it looks right, join the dots.*
* *Don't forget to add a north arrow and scale, address and date.*

Base mapping flowchart

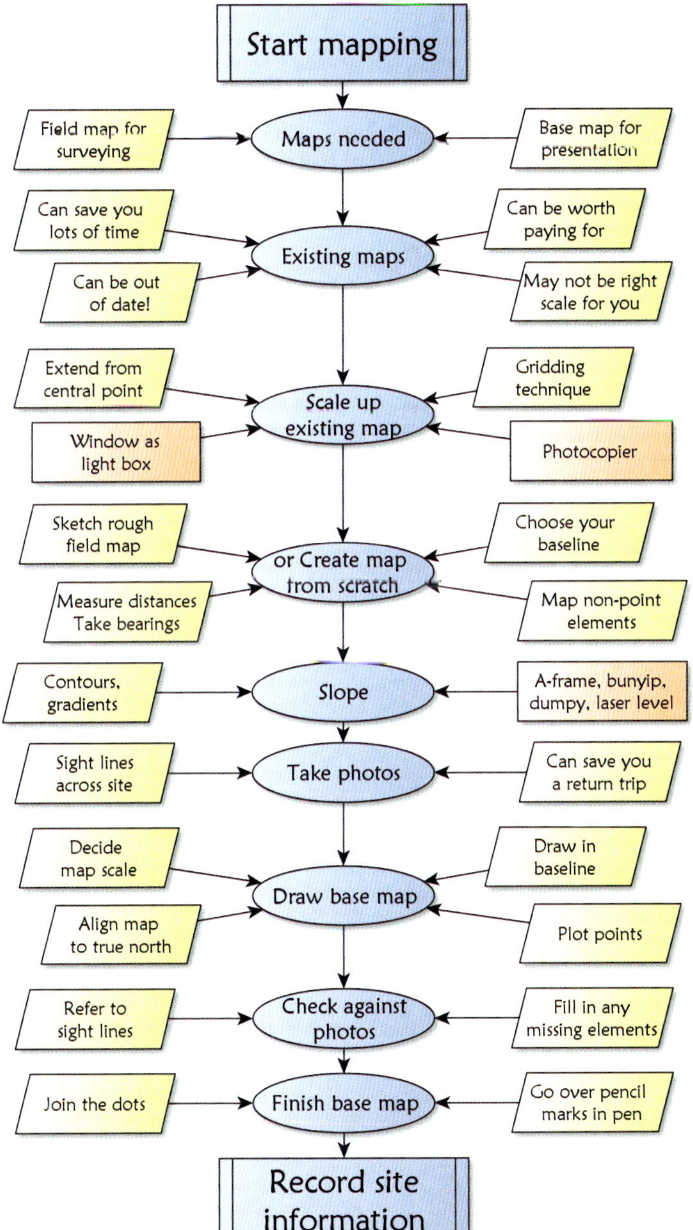

Recording site information

Even if you've been given good maps and photos, perhaps even video, there's so much you'll miss if you don't get onto the site yourself. *The map is not the territory*, so get out there and start using all your senses. And preferably year-round.

Climate and landform

That said, if the eco-system you're designing in is an unfamiliar one, you'll first need to gain an understanding of the local geology and climate. What you may take for granted at home may be completely absent there. We'd be wise to never make any assumptions about what will and won't work in any given place, even in familiar landscapes. Ultimately, the strategies we choose to apply will have to be suited to the local resources and limitations. For instance, in really cold climates, frost might often fill whole valleys and be a more important limiting factor than wind in the planting of orchards, leading to them being located high up on slopes in those places.

Scale of the project

Another factor that will determine your approach is the size of the site. A small garden design will elicit a completely different approach from that for a broadacre farm. While for most of us the former is the most likely place we'll find ourselves working, there's no shortage of farms in desperate need of a redesign. Thankfully, the likes of Darren Doherty and Joel Salatin are showing us better ways under the banner of Regenerative Agriculture and we may all soon find ourselves offering our design services to farms as well. Bill Mollison suggests that the first thing you should do when arriving on site is to get to the highest point and consider the landscape from there. I think it safe to say that if you only have a small garden that he doesn't mean to get up on your roof, but on a broadacre plot this strategy can give you a clear overview of the terrain.

Now, if you've ever tried to dig a pond by hand you'll know how much energy it takes to change the landscape. That's why it's useful when designing broadacre properties to consider ***Yeoman's Keyline Scale of Permanence***.

This scale reminds us of the relative permanence of the main systems we'll be interacting with; the most permanent (unchangable) being at the top of the list. Bill Mollison and David Holmgren suggest an adaptation to P. A. Yeoman's original scale when planning permacultural systems:[9]

1. Climate 2. Landform 3. Water supply 4. Roads 5. Plant systems 6. Microclimates 7. Buildings 8. Subdivisional fences (fields) 9. Soil

Given that climate takes a lot to change (though we're having a good go at it) and the landscape can only be easily modified to a small degree, this means designing water systems first, which define the placement of roads, then trees and so on. Given that water moves across a site away from the highest point and at right angles to contour it's an important location to start from.

Collecting data

Now using your field map or more accurate base map, start collecting data. Rather than cluttering up your map with all of this, you can use overlays† to record the different types of information. Place one onto your base map and trace out the boundary corners and north arrow. Doing this will enable you to reposition it onto the base map accurately later. If you instead use copies of the base map, then start with one of those. Give each a title that identifies the type of information you're recording on it. Even if you use base map copies during your survey, you can still create overlays later from the collected data. Note: your original field versions may be quite rough and not as pretty as those examples shown on the following pages. Some of the following site observations can be mapped. Additional information can be recorded on a separate sheet.¤

Record existing site elements

* *What's the* primary *land use on the site?*
* *By way of a comparison, what is the primary agriculture / traditional crop in this area?*

This information may help you to identify what already grows well here. However, don't allow that to make you any less observant about the specific conditions on this particular site.

† *Alternatively we can use multiple copies of our base map, though these are less versatile than overlays, which can be stacked for easier analysis of information.*

Examine all boundaries (walls, hedges, fences, waterways etc.). Note what they are made of and how well they are being maintained. Make a note to find out who owns each one.

Now add to copies of your map or overlays all the main elements currently found on or making use of the site. Use the ***PASTE*** acronym to help you remember everything:

- *What **Plants** and trees are growing on the site: on the land and in any bodies of water? You might record a transect† across the site as part of this process. Note any fungi you find too – while they're more closely related to the animal kingdom, there is no F in this acronym.*
- *What **Animals** (domesticated and wild mammals, birds, insects, fish etc.) are using the site? Look for any signs of their activities. Don't forget about humans either.*
- *What **Structures** are there (e.g. buildings, greenhouse, shed, paving, pond, dam, terracing, wind turbine pole etc.)? What condition are they in? Are they all being used to their full potential? Any archaeological or sacred sites?*
- *What **Tools** are being used here (e.g. wind turbine, pole lathe, washing line etc.)? What work is done here?*
- *What **Events**‡ take place here (e.g. social: parties, courses, camps – or natural: floods, frost, fire etc.)?*

You may not be able to identify every tree, plant, fungus or animal track. This is another time when photos can be useful. Otherwise, if there's enough to spare, collect a leaf or flower for identification later, either by the client, another gardener, or by referring to a book.

Rain gulley, animal track or both?

† *A path along which one records and counts occurrences of the phenomena of study (e.g. plants, noting each instance). ~ Wikipaedia*

‡ *I've also seen this written as Activities, which makes the acronym PASTA.*

With any plant or tree species, you could also record them using the **DAFOR** framework, noting the relative abundance of species in a particular area. The letters stand for **D**ominant, **A**bundant, **F**requent, **O**ccasional and **R**are. Using this you can quickly describe the overall flora of a site without having to count any individual species. Some experienced botanists also add a 'Missing' category, for those species they are surprised not to find. Remember though that what you see above ground changes through the seasons, and that you may only be able to survey the site at one point in the cycle. Make a note of any questions you have for the client; it helps to have a separate sheet to record these on. Remember also to ask them for any photos they may have of the site taken during other seasons.

Map access points and routes through the site

Using either an overlay or a new copy of the base map, record the access points for people, animals and any vehicles too. Are they adequate for potential site developments and also in good repair? In particular check the state of any bridges or similar structures. Check access roads to the site and the areas they travel through. In fire risk areas, you'll want a safe escape route, not one passing through a potential inferno.

I learned a good lesson about access when choosing our current home. Having first viewed it just a week away from the midwinter solstice, and seeing the house getting sunshine over the hill to the south, I was happy. What I didn't consider until we'd moved in and it had snowed, was that our steep access lane came up the hill from the shady valley where snow and ice are very slow to melt. Ooops!

Coming back to our survey, note what constructed routes (pathways, roads etc.) already run across the site. Are there any additional desire lines? These are paths created by humans, vehicles or animals taking the shortest or easiest route between two places. Here's a great example from one of my favourite books: '*Crap Cycle Lanes*' – go buy a copy, it's for a good cause!

Are there any significant barriers to such flows, such as wildlife corridors into the site being interrupted by busy roads or by waterways? Where do people visit to perform tasks, collect things or to rest? Where are resources brought onto the site, stored or taken away? How does the current layout affect the performance of the site? Are any routes longer than they need to be, or passing through areas causing avoidable disruption?

Identify the different zones on the site

Zoning is all about how energy is being used on the site. We'll start by mapping current patterns of activity and later redesign for greater efficiency. Zones are focused upon main areas of use such as buildings (often called zone 0) and any well-used desire lines, where people move slowly enough to notice what's going on around them. To minimise work we'll later gather the things needing the most attention around these focuses.

Now sketch out the *current* zoning onto your *desire lines and access* map or overlay. Using a different colour for each zone is a great way of differentiating them, like in my example below.

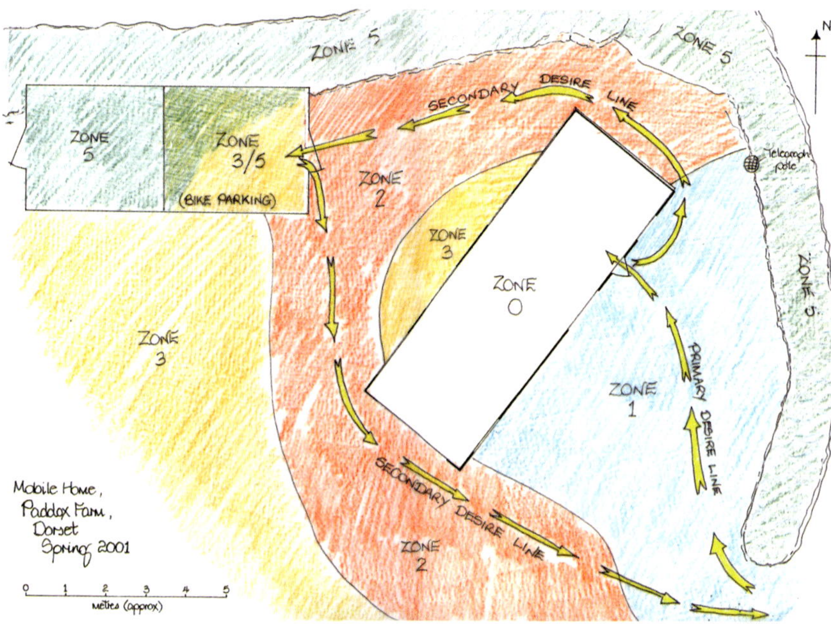

The desire lines around my mobile home shaped my zoning of the space

Conventionally, to define the different zones on a broadacre site we'd follow this rough guide:

> ***Zone 1*** – *Closest to the home, especially any access points (front and back doors), also alongside regularly used paths. Where most intensive gardening and social use occurs.*
>
> ***Zone 2*** – *Gets a bit less attention; may be further away, but still perhaps intensively gardened. May include soft fruit, fruit trees, small animals, etc.*
>
> ***Zone 3*** – *Commercial production (less intensive) e.g grains, main crop vegetables, orchards, agroforestry, fodder crops, pasture, meadow etc.*
>
> ***Zone 4*** – *Managed woodland (often to stabilise steep slopes): timber, coppice, fuel, forage etc.*
>
> ***Zone 5*** – *Places we humans rarely visit (only to observe and to learn). Where wildlife moves freely.*

We'll not necessarily identify all five external zones in this space, especially if the site is relatively small (e.g. an average garden). We'll almost certainly find areas that we can designate as zone 1 and possibly 2, along with zone 5 areas for wildlife; even if these are limited only to drains, gutters and a patch of wild plants. We shouldn't forget that roofs, the tops of trees and inside hedges also act as good zone 5 areas.

With larger sites, especially broadacre farms, we'll also be able to identify zones 3 and 4. Depending upon the site, you may notice that zone 0 (the central focus of activity) isn't a constant, but moves seasonally. This could be the case in a community garden, where the social hub may be a passively warmed greenhouse in the winter, but switch to a shadier spot during the summer. You may also identify more than one zone 0 in regular use on a site, such as a home *and* a workshop.†

In urban areas, regularly used community spaces such as schools and colleges, allotments, libraries and leisure centres may be considered in zones 2, 3 or 4 depending on their proximity. Remember, at this stage we're just identifying the zones as we observe the space *currently* being used. You may well identify opportunities to say, turn a zone 2 into a zone 1, but your job for now is to just record the site's use as it currently stands.

† *David Holmgren calls this Network Analysis in* Hepburn Permaculture Gardens.

While zones are all about conserving energy *on* site, **sectors** are all about the energy coming in from the outside. Often called *wild* energies (as they tend to be beyond our control to do much about outside the site boundaries), these provide us with many opportunities to ***obtain a yield***. Ultimately, all energy flows from a *source* to a *sink* and sometimes our site is in between. The energy source is almost always the sun, even though it may have arrived on site in the form of wind or water flow. Our aim is to harvest, store and cycle energy as much as possible before it's lost again from our system. This 'leaky barrel' analogy illustrates the concept visually.

Left: Energy flows through the site without being accumulated.
Centre: Energy storages are increased, less is leaving the site.
Right: Energy is also cycled around, increasing storages further.

Map the different sectors of the site

Follow the same technique as before to map out the sectors on the site. Given enough overlay/tracing paper, I would use at least two sheets, maybe more. The reason for this is that later on when we do our analysis, we may be considering different combinations of these influences during the placement of each element. We may need to consider the influences of wind and flooding for one element, whereas for another our concerns may be sun and water availability. So ideally use a sheet for each sector, but I'd suggest that if you are to combine sectors onto overlays to save paper, to group them like this:

* *Those sectors that are directional, their angle varying little across the site (e.g. sun, wind etc.).*
* *Those sectors that are topographical: mapped onto specific areas of the site (e.g. frost, flooding etc.).*

Directional sectors

The main examples of these are:

- *The midwinter and midsummer sun sectors.*
- *The directions of both the prevailing and the coldest, most damaging winds. Consider average and gusting speeds, plus changes in moisture content and temperature too.*

Essentially, anything that has a considerable effect over a long distance fits into this category. If you are on a coastal site, warm ocean currents could be considered a directional sector. Winds might also bring industrial pollution from far away.

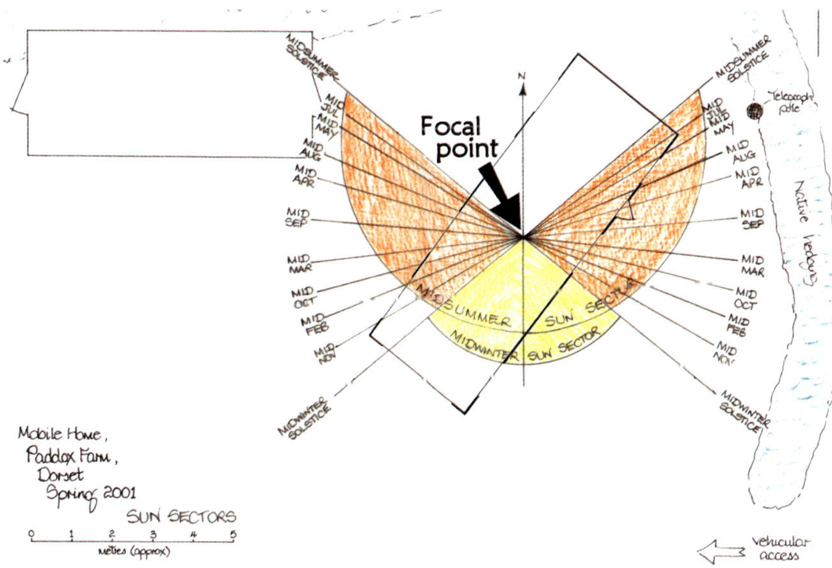

Directional sunrise and sunset angles for midsummer to midwinter with monthly variations shown – for latitude 50°North (London)

Whilst these energies can all be influenced by on-site elements, which may throw shade or slow winds, directional sectors come on to the site from the same bearings regardless of where you are standing. They can be so consistent that, right across Britain, the prevailing wind is considered to come from between south-west and the west. The sun's path is also fairly consistent over a wide area. One important thing to remember is that directional sectors always refer to a focal point which should be indicated in some way. They also need to be aligned to north.

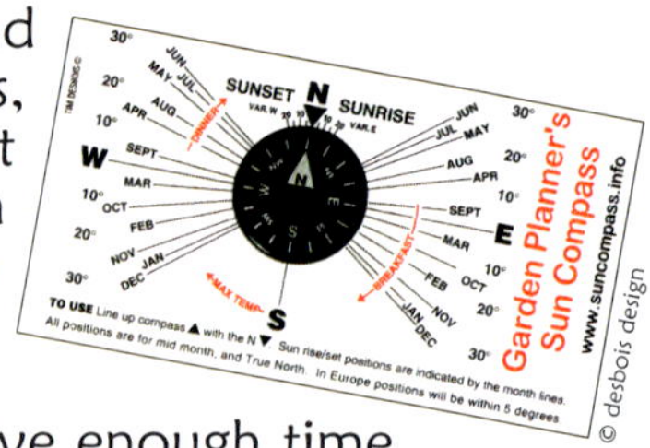

My sun sector overlay was created using a tool called the Sun compass, which shows the changing angles at which the sun rises and sets through the year. It's calibrated to 50° North (e.g. London), but the instructions show you how to recalculate these angles for different latitudes. If you've enough time (seasons) to observe the site, you could create a shade map, showing how this changes across the site from morning to evening and month-by-month. You can place the focus of directional sector overlays anywhere on your base map; align them to north and they will tell you what you need to know in relation to your point of interest. Conveniently, once you've made a sun sector overlay like my previous example, you can use it for any other design sited at approximately the same latitude.

Topographical sectors

The main examples of these are:

* *Areas prone to flooding.*
* *Areas prone to frost/freezing.*

Water and cold damp air reliably flow downhill and collect in pockets or behind obstructions like buildings, hedges, or walls. Ice can build up on access routes where a slope sheds water onto a road or path. The levels of flooding rivers rise up on contour. As a result, these sectors are definable, though their extent can vary with the severity of any weather. We'll come to that shortly. While most directional sectors are represented as 'slices of pie' coming into our point of focus, topographical sectors can be any shape as they're created by the landscape or the structures within it. So remember to 'locate' your overlay on your base map before you start, by marking a few fixed points such as the corners of buildings, This will allow you to quickly align these overlays in future.

Combination sectors

Many sectors we might want to consider though are neither purely directional nor topographical, but a combination of the two. They are formed where a directional sector interacts with a specific feature on the site. Often these lead to ***microclimates*** that give us the variety of niches we appreciate as designers.

Some examples of these combination sectors include:

- * *Wind funnels and sheltered areas.*
- * *Rain shadows of trees, hedges, walls or buildings.*
- * *Shade – remember shadows move through the day and their lengths change through the seasons. (See photos (p77).)*
- * *Nighttime light pollution between buildings/trees etc.*
- * *Any particularly good or bad views.*
- * *Privacy.*
- * *Any neighbouring fire risks (often seasonally inflammable materials upwind of the site) – a big issue in drylands.*

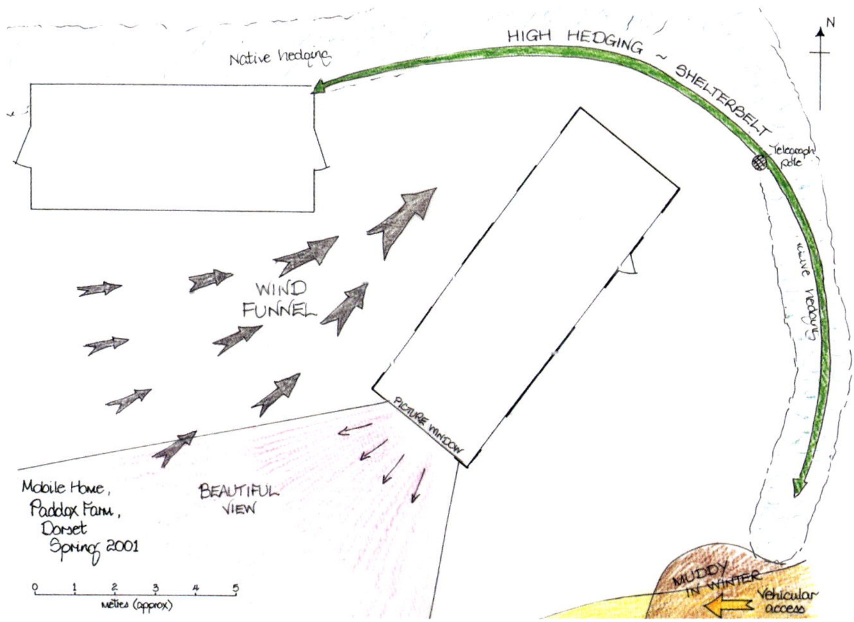

Combination sectors that were affecting my mobile home garden design

In the example above, the view is directional, but very much associated with the picture window of the mobile home as this was where it was most often appreciated. The wind is from the prevailing direction, but the relative positions of the shed and the mobile home are what create the wind funnel effect. The high hedge to the north deflects cold winter winds and creates a nice sheltered microclimate for tender plants in the garden.

Again any overlays you create will first need to be 'located' onto your base map. Other examples of unhealthy human-made sectors that you may need to address are noise, smells, water and airborne pollution such as agricultural chemical spray drift, strong electromagnetic fields, maybe even geopathic stress if you're sensitive to it. Remember to include anything on neighbouring properties that affect the site (e.g. shade from trees, toxic run off from artificial fertilisers or groundwater contamination), and notice anywhere that this is occurring in reverse. Notice too any sectors that need preserving, like wildlife corridors or places of tranquillity where a special connection with nature may be felt. Additionally, look for areas where resources are being focused, such as along garden walls or at keypoints on broadscale sites, where extra water and fertility can be found.

Inclement weather

Whenever it's wet or cold or windy, we usually get indoors as soon as we can. However, it's just that kind of uncomfortable weather that can give us the most important information about a site. One permaculture approach is to ***design for disasters***, and it's important for us to become aware of seasonal weather extremes and their patterns so we can plan for them. Unusual weather events can provide us with insights into these potential issues as they reveal a whole collection of things that we don't normally see, such as:

Heavy rain – Where is water being focussed? How does it flow down slopes? Are there signs of erosion where water is flowing quickly? How does rain flow off roofs? Are there any leaking or missing gutters? Does water infiltrate any buildings? Are there any overflowing drains? Where does water puddle (this can also be a clue to soil compaction)? Are watercourses coloured brown by soil being washed off fields? By the time you've learned the answers to these questions you're likely to be quite wet, but you'll have learned a lot about your site.

Water run off erosion gulley

Strong winds – Where are the windiest areas? Are there any places where litter is blown in circles? Or never settles on the ground? Where are the most sheltered spots? Poles with simple flags (e.g. carrier bags) placed around a site can enable you to observe wind patterns across a large area from a distance.

Drought – When rain is scarce, which areas are most affected by the lack of water? Where is heat and drying out most prevalent? What happens during the first fall of rain after a dry spell – does soil capping prevent the efficient infiltration of water into the soil? Is this problem worse on steeper gradients? Where is moisture retained the longest in the soil (fungi are a big clue to this)?

Frost – Where does frost settle? Is cold air trapped and unable to move down-slope, by hedges, walls or buildings? Could this be remedied? Are there frost-free areas around trees, under hedges, or around buildings? Go out early in the morning to view this – a few hours later the remaining frost may no longer remain in the really cold places, only where shade prevents the morning sun from melting it.

Frost pocket: The downhill flow of cold air off the moorland plateau is interrupted by the building, which is in turn chilled by the trapped air

Snow – The thaw after snow shows us more than we can normally see. Snow melts more quickly on the roofs of heated buildings that are poorly insulated. Capped wells and other underground bodies of water will melt snow more quickly than surrounding ground; a clear circle on a snowy yard is probably an old well that has been concreted over. Other favourable microclimates, such as around buildings, will also thaw snow quickly. Conversely, the chilliest spots will hang onto snow residues the longest. Desire lines are also easy to see in the snow. Where do people and animals prefer to walk? Now you know!

Fire – Hopefully you'll never experience a wildfire, by all accounts it's a pretty terrifying thing. If the site is in a fire risk area then designing to protect the site against it is a priority (remember designing for disasters?). Instead of waiting for one, find out about the history of wildfires in the area and their patterns. Radiant heat is the most destructive aspect of fire and burns from a considerable distance, even a small campfire can force a retreat. The main things to look for on site are:

* *Inflammable plants and trees; conifers and eucalypts (high resin content trees) burn particularly fiercely. Do prevailing winds blow on to the site from that direction?*
* *Is the site on a slope? Fire travels very quickly uphill and is fiercest on ridges, which are usually the driest areas.*
* *Where are access roads to the site routed?*
* *Is there an emergency on-site gravity-fed water supply?*
* *What are the buildings made of? Are they designed with fire-protection in mind (e.g. white painted, with door and window screens, simple roof shapes and screened undershot guttering that doesn't collect hot ashes, etc.)?*

Solar elevation angle

Doing this will help you find different microclimates on site, together with opportunities to harvest incoming energies and areas of the site that need protection from them. Remember that some sectors have vertical components too and can vary in strength with altitude. Low winter sun can be blocked by tall trees and buildings, so it's useful to be able to approximate and record their heights, to estimate seasonal shade patterns. The Sun compass also allows you to estimate this changing shade using the angles printed on the back (see below).

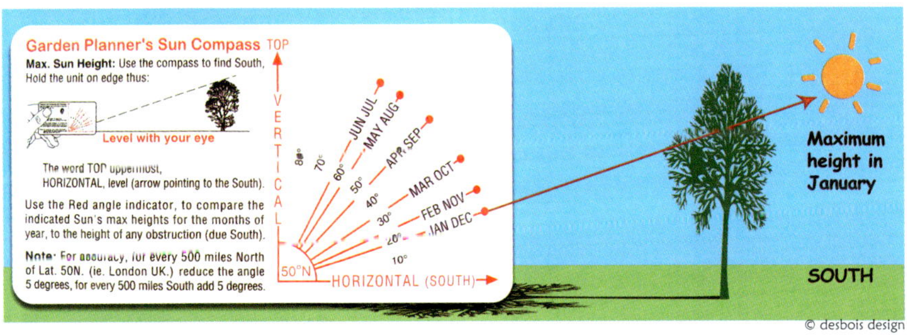

Estimating heights

You may be concerned about the shade a building or tree will throw in the winter, or want to know the protected distance offered by a windbreak, so it's useful to be able to estimate the heights of tall objects. The simplest method involves comparing the height of say a tree, with a person standing beneath it. This can be pretty accurate where the tree is only a few times the height of the person. An adaptation of this is to compare the length of your own shadow with the tall object. Pacing works pretty well for this, though if the ground is uneven or of changing slope, this can distort the result. In case these prove insufficient, I've included a slightly more detailed but accurate method online.¤

Record water across the site

- Are there any open bodies of water present on the site (ponds, lakes, reservoirs etc.)? Any opportunities in the landscape for creating more (especially in clay soils)?
- Any water passing through the site (streams, rivers etc.)?
- Are there any drainage ditches or earthworks?
- Any signs of wells/boreholes or monitoring of aquifer levels?
- Estimate the surface area of any roofs and note which of them have guttering and to where they drain rainwater.
- How much water storage is there currently on site for different uses (drinking, washing, flushing, irrigation etc.)?
- How much is stored in tanks, butts etc.? How long do these last during drought? – a question for the client.
- How are water systems connected on site? What water cycling is currently taking place?
- Is grey or black water being treated before leaving the site?

Take at least one soil sample

What do the locally abundant wild plants tell you about the soil? Before digging a hole, use the indicator plants chart¤ to determine what kind of soil these plants like to grow in.

Choose sites to survey away from compacted areas like paths and roads

To see how accurate this is, get a copy of the Biological and Soil Monitoring Chart¤ and a spade or two. Then:

- Find a suitable site, away from compacted areas like access points or pathways.
- Start filling in the chart, starting with column 1 for site 1.
- Sketch a cross section and mark the location on it.
- First make a note of the current land use there.
- Determine the degree of plant diversity – to do this properly you might use a one metre square frame (or a hula hoop!) and study just what is inside. Record or sample plants.
- Record any signs of insect, spider and soil life activity (worm casts, nibbled leaves, webs etc).
- Now dig your soil profile (a square hole a spade's width on each side and a spade blade deep). How easy is it to dig? – This is a clue to the degree of compaction!
- Carefully remove an additional downward 'slice' to one side. Sandwiching it with a second spade can help to lift it out. Observe the different soil layers, roots and soil life.
- Smear a little soil onto the chart (in the colour box).
- Taking a representative sample of topsoil use the 'How to test soil texture' flowchart¤ and follow the instructions until you have determined your soil type.
- To check this, also perform the jar test¤ on a further sample from the same place. This picture clearly shows the layer of settled sand that falls straight to the bottom, with silt quickly settling out over the next 30 minutes or so on top of it. The paler clay is still in suspension on the top and may be for many days or even weeks in the case of really fine clays.

If your site has a variety of different microclimates and land uses, it's worth surveying the soil in a few places and then comparing your findings. What do these observations tell you about the underlying geology of this area? The base rock will determine the soil type and thus the basic growing conditions.†

Identify the site's remaining limiting factors

Our survey may have already identified some of the site's key limiting factors: perhaps excessively shady or boggy areas, very heavy or light soil, or crops regularly grazed off by insects or wild animals? Some of these factors we may seek to modify, others such as altitude we are going to have to accept and seek to discover the gifts they offer. Our role as designer is to identify *key* limiting factors, and then to design strategies to overcome them. Sometimes, by removing one limitation, the landscape will change dramatically, like removing grazing animals from a landscape to permit the regrowth of forest. Returning to our leaky barrel analogy, an effective strategy might be as simple as plugging some of those wasteful leaks. A quick look around many sites will quickly identify the tragic loss of energy and resources such as:

Chris Dixon in his woodland, which regenerated after sheep were removed

- * *Heat escaping from buildings.*
- * *Fertility being washed out of the soil.*
- * *Water leaving the site before being fully utilised.*
- * *Crops being left to rot (most commonly under trees).*
- * *High maintenance, low output systems (e.g. most lawns).*
- * *Vandalism.*

You may also identify other opportunities being wasted like:

- * *Workers having insufficient to do, or being wasted on low value tasks. Volunteer help not being made use of.*
- * *Free or cheap local resources, not being collected or utilised.*

† The Living Landscape *by Patrick Whitefield has an excellent chapter on this.*

In addition there may be other 'non-physical' limiting factors to consider, such as:

* *Legislation (e.g. planning, conservation etc.).*
* *Ownership.*
* *Cultural issues (including the reactions of neighbours).*

Explore them as part of your client interview process. A mind map of limiting factors in the online resources ¤ will help you identify others that relate to your site.

These non-physical limitations can be considered sectors too, as legislation can change unexpectedly making current site systems untenable. One example being a successful juicing business, built up over many years, by buying fruit from abandoned, unsprayed orchards. Then one day their main customer decided that everything they sold must be certified organic. This seemingly good decision meant the farm had to find a new source of fruit and abandon those old orchards to a possible grubbing out.

Map any site utilities

Are any mains services (gas, electricity, sewage, water, phone etc.) already being supplied to the site? Where are they, above or below ground? If the latter isn't clear, then utility company maps can help identify the routes of buried pipes and cables. This will help you to avoid accidentally damaging any during excavations. In which parts of the site are such utilities available or easily connected up to?

Where is the nearest settlement?

Where can resources be obtained for site development (e.g. building materials)? Where are the nearest shops for food? Where can medical help be obtained in an emergency?

Identify any free or cheap resources available

Finally we need to record the valuable resources available on or close to the site. These may include ***natural resources*** such as timber that can be harvested sustainably. Space is a resource that many people don't have, so any buildings standing unused offer the opportunity for new ventures, including renting them out to other businesses or as a venue for running courses.

There may be specialist machinery available nearby that could be leased, and local people with specific expertise that can be employed to help develop aspects of the site.

Often items considered to be junk by most people can be put to ingenious use. Remember, ***the problem is the solution*** – it's only the way we look at things that makes them a problem or an asset. Here are a few examples:

* *Tyres being made into rodent-proof wormeries (of course, if you have enough, you could even build your own Earthship dwelling!).*
* *Old baths being turned into small ponds and all kinds of other large containers used to store rainwater.*
* *Pallets made into compost bins.*
* *Patio doors, recycled plumbing bits and old copper pipe being made into very efficient solar hot water panels.*
* *Double glazing units thrown out by window companies made into greenhouses/bioshelters.*
* *Large plastic bottles (from cooking oil, water etc.) being made into cloches.*

When travelling to and from the site, make a note of any local businesses that may be throwing away useful items, such as any of the above and also:

* *Cardboard (for mulching).*
* *Green waste (for making compost).*
* *Manure (need I explain this one?).*
* *Glass bottles (make good raised bed-edging).*
* *Waste timber (for building garden structures).*

And so on... ***We are limited only by our imagination***! So make a note of everything that's available on the site, or that you've noticed nearby. Any of it may become an integral part of your design.

Creating base map overlays

Once you're back at home again, you can turn your rough field data into bigger and clearer overlays. As previously mentioned, if you have plenty of tracing paper it will give you the most flexibility in making decisions later. Remember. if you need to group different data onto single overlays, then try to keep directional and topographical sectors on separate sheets.

Place each fresh overlay sheet in turn over the base map and unless it's to record only directional information, first mark on several points to locate it (corners of boundaries or buildings are ideal). Then transfer across, a sheet at a time, all of the information you've gathered from your rough field maps or overlays. I find when doing this that a good dark pen is useful for defining boundaries and that using plenty of colour shading helps to show up different areas, such as zones.

Remember that directional sectors such as the sun's path, don't need to be the full size of the map. You may wish to centre them on different points of interest when you're later making decisions, so a little smaller can be more convenient.

Keeping an observation diary

Earlier we noted that it's best if we can observe a site for a full cycle before making any 'hard-to-reverse' decisions. Of course, for any land-based design that means a whole year and most of us are far too impatient to wait that long before starting. However, by making only small changes in the first year (e.g. growing annual vegetables) while continuing to observe the site, we can enjoy some progress and yet reduce the risk of making any really stupid mistakes. We can even make some early plantings part of our observations. If for instance we wish to plant an orchard in a year's time, but are still uncertain about the best site to choose, we could buy some trees and plant a few of them in each of the sites we are considering. That way, the trees themselves can report back to us about how happy they are in each place.

It doesn't really matter how we record our observations, as long as we do something. As long as we keep our handwriting legible, any record is better than relying on memory, something I've discovered the hard way when examining unlabeled packets of seed each spring! Yet there's no point in setting yourself up to document lots of information every day if it soon feels like a chore. Just record what interests you (presumably what you're designing) and make it easy to do or you won't bother.

A simple diary format might consist of a grid of boxes into which we record physical conditions like temperature, rainfall, perhaps even river levels etc. – any factors that relate to what you might be designing for on your site. I keep a simple weather log, which has columns for three years worth of observations, allowing me to notice any long-term patterns. And because it's a simple format, I rarely fail to fill it in.

With digital cameras even being built into phones these days, we have less and less excuse to not make some kind of record of what is going on around us. Photos are of course a great way of recording visual information and can save us plenty of time that would once have been spent drawing or writing notes. Recording shade patterns across a garden through the seasons is particularly well suited to this technique.

Make a note on your calendar to take a photograph from say an upstairs window early morning, noon and evening at the beginning of each month (when the sun is out of course). The resulting set of photos you'll find to be very revealing about microclimates through the year.

A comparison of two photographic shade maps from one of my former gardens

Alternatively, you might choose instead to write about or draw, paint or photograph what you see in nature on any given day, like Janet Marsh[10] and Helen Moore[11] have both done so beautifully. This kind of diary will be more detailed, but have less 'raw data' to work with later. Another useful record to keep is a phenological diary. Phenology is the study of plant and animal life cycles and is useful because it has allowed us to track climate change by observing nature's responses to it, for instance the timings of the appearance of buds, flowers and fruit and the migrations of birds and animals. While such a diary can be a useful observation tool for our own site, there is now a whole network of observers across Britain that allows the UK Phenology Network to build up a countrywide picture of these patterns. Your own observations are a valuable contribution to this and it's very simple to sign up on the Nature's Calendar website.[12]

Summary

The key things to remember about recording site details are:

Recording site information

* *Use copies of your base map or overlays to record spatial information and a separate sheet for other observations.*
* *Record what is happening on the site now.*
* *What is the primary use of the site? What else can you observe here? Use PASTE and DAFOR to help you.*
* *Keep a diary if you can observe over a period of time.*

Zones and sectors

* *Map the current zones, access points and desire lines (human and animal) through the site.*
* *Observe the sectors, both directional and topographical, ideally using different maps or overlays to record them.*
* *Inclement weather can show you things that you don't normally get the chance to see.*

Soil and water

* *Take soil samples to build up a picture across the site.*
* *What do the common plants here tell you about the soil?*
* *Record any significant bodies of water.*
* *Identify the amount of water storage in ponds, tanks etc. What clues are there to the current level of water use?*
* *How is water being treated before leaving the site?*

Resources and limiting factors

* *Record any valuable resources on or near the site.*
* *Identify the key limiting factors on the site.*
* *Where is the nearest settlement?*
* *Map any utilities already on the site.*

Recording site information flowchart

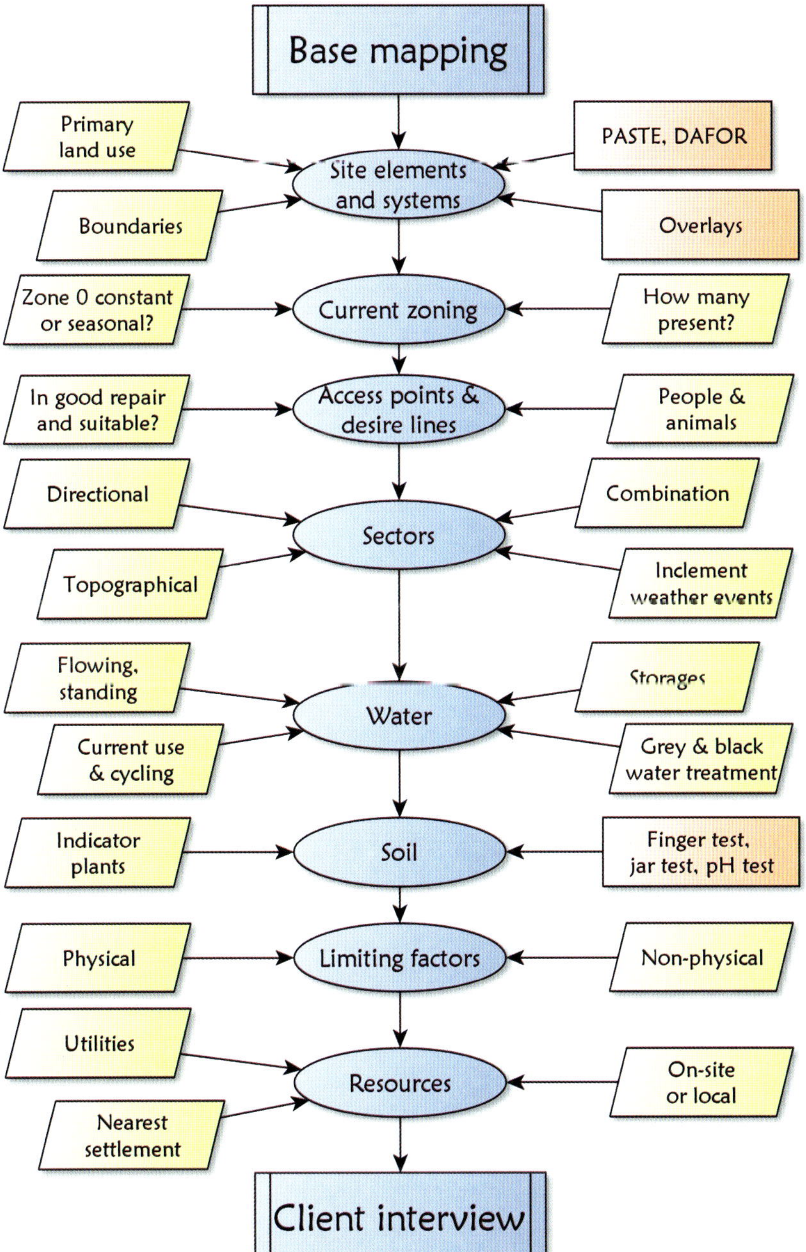

The Client Interview

This is the part where we identify why we are carrying out a redesign of the site and for whom. Ideally we would go through the following process with each client (everyone involved ought to be interviewed to some degree), either one at a time, or together. Find out who all the clients are in advance. Sometimes, they will include both adults and children, though one or two may guide the process more than the others. You may also need to take into consideration the needs of specific animals (even if you cannot question them directly!).

Sometimes, the client list will include you. If so it can be helpful to get someone else to ask you these questions and make notes for you. Not all of them will always be relevant and, depending on the circumstances, some will be more important than others. So give yourself sufficient time to prepare your list of questions before meeting the client. To help you I've included a client interview worksheet¤ as part of the additional online resources, which you can adapt for your own needs.

SMART goals

Out of this interview you need to identify a set of goals, a way by which you can measure the success of your design over time. In order to do this, those goals should have ***SMART*** qualities; in other words they need to be ***S***pecific, ***M***easurable, ***A***greed upon, ***R***ealistic and ***T***ime-bound. Well, these were the original words associated with SMART, though you'll find other versions of the acronym these days too. The blackboard (right) shows a different interpretation.

The interview

This interview process alone should work fine for up to a few clients. Once groups get bigger though, some additional tools might be used to ensure that everyone gets heard. We'll consider some of these different methods shortly, but first let's look at a typical client questionnaire. You'll mostly likely have a limited amount of time with the client(s), but if you can, don't hurry them on any particular question. Often the most important answers will be the ones you receive first. However, given enough uninterrupted time to ponder, some real gems can emerge. Adapt the following common questions to suit the time available and the particular circumstances. You might even ask those that could benefit from some deeper consideration in a pre-meeting questionnaire.

How do the clients feel about current arrangements?

Start by asking what is already working well for them and what clearly isn't. This will tell you what you will be designing around. Things to consider in a garden might include:

- *The overall layout – how well is it meeting their needs?*
- *Access points, paths and general accessibility.*
- *Structures – sheds, pergolas, water features etc.*
- *Trees and plants – which are appreciated/disliked?*
- *Site maintenance – and do they currently need help?*

What are the client's needs and wants?

What are their currently unmet needs? You might use PASTE again to help them think about this. Divide these into:

- *The priorities that they feel they must have?*
- *The additional things it would be nice to have?*

What are the client's values and vision?

What is most important to them? You must design taking into account the client's values (e.g. a vegetarian might not be too impressed if you hand them a livestock-based design). What is their vision for the site/project? Get as much detail as you can. Let them draw, paint or make a collage if they wish.

Not everyone has great clarity about this though, they may only know what they don't like about the current arrangement. If your clients feel a bit stuck about this, show them a collection of images cut from magazines (gardening, nature etc.). You might want to take some along just in case.

Lifestyle questions

It may also help you to know some of the following:

* *Their occupations – how do they earn their living(s)?*
* *Income – is any earned by selling products from the site? If so, are any resources are being imported onto the site to enable the making/growing of these products?*
* *How much time do they spend on site? Does this vary with time of day or time of year? How much time will they have free to implement/maintain the design?*
* *Transport – how do they currently get around and where are their regular destinations? Why do they make these journeys?*
* *Hobbies and interests – what do they like to do? How might they want to use the site? Do they like gardening?*
* *Their eating habits – how local, seasonal, organic etc.?*
* *Do they already grow food for themselves?*
* *Frequent visitors – are there any other people that use the site often, that also need to be considered?*
* *Do they regularly entertain friends on the site?*
* *Does anyone (residents or frequent visitors) have any allergies or anything similar that you should know about?*

What are their personal limiting factors?

What's stopping the client(s) from realising their vision? Your task as designer will be to identify their key limiting factors, but for now, just help them with ideas if they get stuck, and write everything down. Limiting factors¤ might include:

* *Physical issues (e.g. poor health or mobility, little energy).*
* *Emotional issues (e.g. lack of confidence, motivation).*
* *Mental issues (e.g. lack of imagination, understanding).*
* *Lack of money, time, skills etc.*

What personal resources do they have?

Like limiting factors, resources can be either physical (e.g. being able-bodied) or non-physical (e.g. being patient). Most things that could be limits, might instead be resources, for instance:

* *Vision/inspiration/experience.*
* *A good network.*
* *Time, energy, skills, etc.*

What is their financial budget for the design?

How much money is available for both the implementation and for ongoing maintenance? You should expect to invest 80% of your energy into creating a system that takes only 20% to maintain. How this works in practice is down to your own judgement, as implementation costs end with the completion of that phase, whereas maintenance costs are ongoing. Remind clients that the availability of funds may determine the speed at which the project can proceed. The same 80/20 principle also applies to time invested, even if in the form of paid labour. Don't get caught up with the numbers, just remember the general concept. Finally, remember that intangible resources such as a good network[†] can be just as important elements in a system as any physical ones.

Site-related questions

It can be really helpful to walk the site with the client(s) if possible when considering at least some of the following:

* *How long have they lived on site and how long would they like to stay? What is their security of tenure? Do they own the site or rent it, if so what lease do they have?*
* *Are there any security issues with the site (e.g. previous burglaries or vandalism)? If the site is locked up at times, what are the access issues? Who has keys etc.?*
* *Who owns and is responsible for maintaining site boundaries? Are there problems relating to their current state of repair?*
* *What rights of way exist through the site?*
* *What do their neighbours do? Are there any problems as a result of their activities; poisonous run-off, spray drift etc.?*

† *This became clear to me a couple of years ago when I helped on a four-day house build, where thirty skilled and enthusiastic friends came to help the family.*

* *What energy sources and resources are currently being used on site and how much? Where are underground utilities routed?*
* *Are there any energy or resource leaks that you missed during your site survey? Where are they 'bleeding'?*
* *How long do storages last (e.g. water in drought, food when snowed in, energy in batteries if relevant etc.)?*
* *Are any outputs currently going unused (e.g. waste heat, orchard produce)?*
* *Are there any animals (e.g. pets/livestock) that also use the site?*
* *Do they have photos of the site in other seasons?*

Any other site related questions?

Don't forget to ask those extra questions that you noted during your site survey. Things like the local availability of resources, including skilled people, any previous use of the site, etc. Find out what they know about planning. Are there any restrictions on what can be built or planted? Does any building or tree have a preservation order on it? Is any of the site in a National Park or listed as an SSSI†/nature reserve? Is there anything in the local authority's development plan that might affect the site? Your client(s) may not have the answers themselves, but they may be able to point you in the right direction.

What is their timescale for the design?

Identify how quickly they would like to achieve their vision. Is it a short-term fix that needs to be up and running in a few months, or do they have a much longer-term view? Perhaps they have both and are looking for a phased design? You'll need to identify their ideal timescale(s) to create SMART goals.

Personal details

Make a note of the interview date and your clients' details: their names, ages (especially children) and contact details: postal address including postcode,‡ phone number(s), email, etc.

Anything else?

Don't forget to ask them that all-important final question...

† *Site of Special Scientific Interest – there are around 4,000 in England, covering around 7% of the land area, so this is definitely something to check. If the client doesn't know you can look this up on Natural England's website.*

‡ *GPS co-ordinates and map references are also useful for finding online maps.*

Working with multiple clients or community groups

When working with larger client groups it is important to ensure that everyone feels heard. Ultimately, the success of a community project depends heavily upon the degree of ownership a group feels over the end product. I've seen well-meant projects, such as community orchards, created with little or no consultation with the local residents. Inevitably those projects suffer vandalism, because those carrying out the damage have no connection to this thing that 'just landed' in their neighbourhood.

Here's a small selection of tools that have been developed to help groups successfully make decisions, any of which may prove useful in particular circumstances. If you plan to work with groups of more than just a few clients, then I recommend that you investigate at least one of the following methods in more detail than I have room for here. I provide just a brief overview of each below, but each is well documented either in books, on the Internet, or both. When working with groups in this way, you might be offering a set of questions for the group to go away and consider their answers to, or be stepping into the role of an outside facilitator in their process. However, if you are considering taking on the latter role, I would certainly suggest that you get some training first.

Small community processes

Any of the methods previously offered to assist design teams in working together could also be useful to small groups of clients in establishing a collective vision and set of priorities. Two additional processes follow that take up more time, but offer a way to ensure equal participation in any key decision making. These work well for up to 10 or 15 people, maybe more, but as the group size increases, the time needed for the process may expand accordingly.

The Holistic goal-setting mandala

I learned this technique from New Zealand permaculture designer Robina McCurdy. She calls it:

> *"An empowered visioning tool for group alignment, to form collective goals and action plans in all facets of community life. As a fully participatory technique, it recognises, honours and encourages each person's gifts, skills, vulnerabilities and strengths. Through a specific method of communication and contribution, everyone is equally involved and valued. This tool enables a community to define what is important, where members/stake-holders have similarities and differences in what they stand for, and what they want to develop, and how the group/project can move forward from a broader vision to implementing details."*

This technique often uses an external (neutral) facilitator. If this sounds like it would be of use in your process, seek out her counsel through EarthCare Education Aotearoa.

Consensus decision-making

This process has several key objectives:

* *Inclusion – as many as possible should be involved in the consensus decision-making process.*
* *Participation – to actively solicit the input and involvement of all decision-makers.*
* *Cooperation – the group aims to reach decisions that everyone is happy with, rather than opting for majority decisions that could alienate those who opposed them.*
* *Equality – the opinions of everyone involved are considered to be of equal importance. Everyone can present, amend or block proposals.*
* *Solution-oriented – common agreements are identified and agreements are achieved through compromise, etc.*

Agreement by consensus can take considerably longer than by majority, but tends to result in stronger community ties. Over time groups become more skilled and the process often becomes quicker. The flexibility of participants and wording any proposal in a way that everyone involved can be happy with it, seem to be the keys to making this process work well.

Small to medium sized group processes

Open space technology (OST)

Created to enable groups to deal with complex issues in a short space of time, OST has been successfully used by thousands of organisations in 134 countries. It has been used to organise meetings for as little as five and up to two thousand people. Having noticed that the coffee breaks were the most productive part of one conference, its originator Harrison Owen set out to recreate a whole process around this. The Open Space element is a large circle in which everything takes place. Participants can write up questions and post them on a bulletin board for everyone else to consider. These issues are then placed on a space/time matrix, becoming the agenda. People attend the sessions they are most interested in, which may take place around the central space or in other breakout areas. There are four principles of OST; Whoever comes are the right people, Whatever happens is the only thing that could have, Whenever it starts is the right time, When it's over it's over. There is just one law; the '*Law of two feet*', which means people can move from session to session if at any time they feel that they are no longer learning or contributing.

The World café

Devised to host 'conversations that matter', the World Café can be a useful process for finding out what matters most to a group of people. Ideal for small or larger communities, a space is laid out with café style tables, each focussing on a particular question. The whole process is guided by seven core principles; Set the context, Create hospitable space, Explore questions that matter, Encourage everyone's contribution, Connect diverse perspectives, Listen together for insights and Share collective discoveries. Similar to OST, people are free to move from table to table to share ideas and this is where valuable cross-pollination emerges. Unlike OST, the theme for the event is often chosen in advance rather than set by the group. Words, images and colour are used to capture participant's ideas and expressions on large sheets of paper. These are posted on walls to enable all to see what is being discussed. This documentation also later serves as the group's memory and enables subsequent sharing with others.

Wider community planning

Transition culture

Perhaps already the biggest social experiment in history, the idea of Transition was birthed by permaculture students from Kinsale in Southern Eire under the tutelage of Rob Hopkins. The Energy Descent Action Plan (EDAP) they produced for their small town in 2005 inspired a global movement that now offers a vision of local resilience in the face of climate change and peak oil. By approaching our impending energy-descent from a positive and creative perspective, Transition provides a model for communities to plan a gentle and enjoyable journey into the future. Transition Culture is simply permaculture being applied to communities. You may already be involved.

Planning for real®

This is a process for involving local people in developing their locality. It focuses around a table-sized model or plan, ideally created by members of the community. The model is then taken around to locations like shopping areas and community centres, and passers-by quizzed about their ideas and opinions. People consider their regular journeys, what they most like and use, and what things they think might improve the area. The information is gathered onto the model or plan as cards and flags, with options for others to agree or disagree with previous suggestions. Because everybody focuses on the model, it helps to avoid direct confrontation between individuals. Variables such as age group, gender and home locations of participants are taken to ensure equal involvement in the process.

Participatory rural appraisal

This is an approach that has emerged from NGOs working in international development. It consists of a large toolbox of participatory techniques, seeking to enable people, especially the financially poor, to take back much more control over their lives. The techniques can be divided into four categories: those to assist group dynamics, those for gathering information, those based around discussion, and those more imagination-based. One of the key ideas is to avoid writing as far as possible. This is in order to prevent anyone being excluded. Instead, the use of pictures, symbols, and physical objects are encouraged.

Case studies

(some more informal methods that have also worked well)

Community space agreement: Permaculture Association CEO Andy Goldring was invited to help facilitate a design process at a low-impact intentional community. One point of contention was the inability of the residents to agree on the size of the proposed village green. Andy took them outside and stood them in a circle, posing the question "Is this big enough?" Clearly nobody thought it was, so he gave them each a stick and asked them to walk outwards until they thought it right. After a certain amount of negotiation, the sticks were planted in the ground to define the boundary and an 18-month process swiftly brought to completion. Andy is a skilled facilitator, but tools like this, along with the confidence to take charge of the process can be a valuable, timesaving skill.

Office redesign: Janey Williams was put in charge of an office redesign where she worked. Few people were happy with the space as it was, but there was reluctance from staff to share their feelings in a formal way. By simply chatting to everyone during coffee breaks, Janey was able to learn about nearly everyone's preferences and grievances. In such a relaxed and less public atmosphere, people are more likely to share how they feel. The redesign was a great success as she was able to meet most of her fellow workers individual needs.

Martial arts club: When learning Aikido a few years ago I set out to create a design to maximise my learning opportunities. It quickly became clear to me that in order to do this, I had to ensure that all my fellow students' and Senseis' (teachers') needs were well met too. Without anyone to teach me or to train with, I couldn't learn. Having made a list of questions for my fellow club members, I gathered opinions on a different subject each training evening (twice weekly), again in an informal way. Because we trained so often, I was able to ask my colleagues about a wide range of issues a little bit at a time, avoiding the need to pin everyone down to a long interview process.

Summary

The key things to remember about the client interview are:

Preparing for the client interview

* *If working for others, identify all the clients with an interest in the design. Try to arrange to speak to them all.*
* *If you are both the designer and the client, find someone else to interview you if possible.*
* *Give yourself enough time to prepare your questions, so that you make the most of the time spent with the client.*
* *Remember that you'll benefit from setting SMART goals, so ask questions that get you clear answers.*
* *Try to conduct at least part of the interview on the site, so that you can ask site-related questions in context.*

The client interview

* *What do they already like? What don't they like?*
* *What are their unmet needs and wants?*
* *What are their values? What is their vision?*
* *What are their personal resources and limiting factors?*
* *What is their timescale for the design?*
* *Ask any site-related questions that you have.*
* *Gather their lifestyle details.*
* *Make sure you give them a chance to add anything else.*

Working with multiple clients

* *The methods offered to help the design team work well together can be just as useful for small client groups.*
* *For really large groups, use one of the suggested processes to ensure everyone can be heard.*
* *The group dynamic and the time available will determine which process is the most appropriate in each case.*
* *Simple informal interviews are sometimes the best option.*

Client interview flowchart

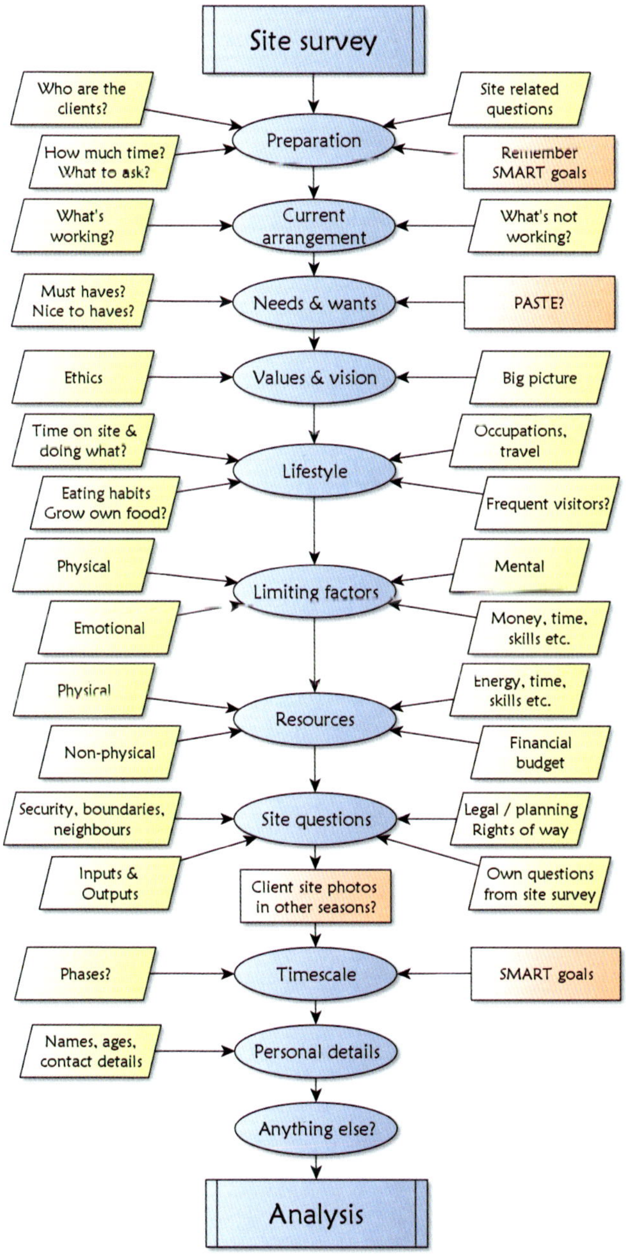

Analysis

"Permaculture design is a system of assembling conceptual, material and strategic components in a pattern which functions to benefit life in all its forms."

Bill Mollison

Sounds good, but *how*? Don't worry, we're going to learn that now, one step at a time.

I feel that this phase of the design process, represented by just the 'A' of ***SADIM***, or the 'E' (evaluation) of ***OBREDIMET***, should receive more of our attention than a single letter infers. Through analysis, our challenge is to design and establish a system where the energy needed to maintain it decreases over time, while its outputs increase.

Remember we aim to invest as much as 80% of our time and energy in establishing it, so we only need 20% to maintain it. A low maintenance or high efficiency design should always be an intended outcome, not just an occasionally chosen function.

In this section I'm going to offer you a collection of design tools that work for me. I think they'll work for you too, though at the end of the day it doesn't matter what tools you use, as long as they work. Ultimately, making decisions can be done in many ways, but the effectiveness of our design usually relates to the amount of time we spend on our analysis.

So, having gathered all that lovely information, we'll now play with it, identify the key functions of our design, and define some SMART goals to guide us. We'll then get out all our ideas (even the seemingly crazy ones) and start narrowing them down to find the best of them. We'll use a few thinking tools to help us to do this, starting with the ***permaculture ethics***. Then we'll see how well our best options integrate together into an effective system and decide the best placements for them in our overall design. Part of this process will involve us thinking about the principles of ecology, which are simply just the application of successful natural patterns.

Identifying functions

This is the stage where we identify the focus of our design. Every design must have a purpose; what is the point of ours? We'll take what we have learned from the client interview(s) and the site survey, and use that information to determine what functions we need to address (there may be many, but some will be more of a priority than others).

When we visit somewhere new, we often start redesigning that space in our minds, based upon things we've seen in other places, things that we'd prefer ourselves. This redesign usually takes the form of the imaginary placement of objects, be they plants, animals, structures etc. While aspects of this initial assessment may turn out to be good, each new space has its own unique set of circumstances and is best approached as such. George Carlin coined a saying: '*Vujà dé*' – the feeling that we have never been in this situation before – and that is the best way to approach each new design.

The process

It's not uncommon that while we're surveying a site for the first time, we'll also have some ideas. So let's get those pieces of paper out of our pockets again. We need to work backwards from our initial insights and identify why we thought those things would be a good choice. What were we aiming to achieve by their inclusion? Was it a need for a degree of food or energy self-sufficiency? Or for protection from strong winds? Did we want to attract wildlife? Or just have a beautiful place to relax? These questions define our design.

For instance if we thought an apple tree would be appropriate, why were we choosing it? Was it because it would feed us? Or stabilise the soil? Or provide an income? Give us shade? Provide good forage for bees? Or just look beautiful when it blossoms in the spring?

Stepping back to identify the functions we want is an important stage in choosing the very best way to achieve them.

Functions, systems and elements

Still confused about the difference between functions, systems and elements? Maybe this will help: elements are individual things that make up a system when connections exist between them. Several systems can also connect up to create bigger systems. Either elements on their own, or complete systems, perform (one or more) functions. A function is *what you want to achieve* and the system or element is *the means by which you achieve it*. Here are some examples:

Function	***System***	***Elements***
Windbreak	Hedge	Hazel, elder, ash, bramble etc.
	Fence	Posts, concrete, wooden slats etc.
Irrigation	Roofwater harvesting	Gutter, diverter, downpipe, water butt or tank, pond
	Earthworks	Dams, sluices, gulleys, trees etc.
Soil improvement	Slope stabilisation	Terracing, swales, gabions, net and pan, trees
	Nutrient cycling	Composting, green manures, liquid feeds, mycorrhiza, treebog
	Mulching	Cardboard, bark, compost, straw, wood chip etc.
Food production	Veg garden (plants)	Carrots, potatoes, onions etc.
	Veg garden (structure)	Raised beds, composting system, greenhouse, hand tools
	Orchard	Apple, plum, pear trees etc.

Note that while Hazel has appeared here as an element in a hedge system to provide wind protection, it could also have been listed as an element fulfilling the functions of soil improvement or food production.

Remember, 'adaptable' (multi-functional), resilient (multiple elements for each important function) and 'high efficiency' (least work for greatest effect) are always design *principles* and don't need to be considered as separate *functions*.

Identifying the key functions

Looking through your notes from the client interview(s), you should be able to identify what their key unmet needs are, along with the less important things that they would like too. Your challenge is to integrate their own needs with the needs of the environment and still give them everything they ask for.

In addition you'll hopefully have identified from both your site survey and client interviews, what the current key limiting factors are. These may be associated with energy or resource leaks from the site, such as soil erosion, water run-off, high maintenance costs, badly insulated buildings, wasted volunteer help etc. Plugging these leaks (soil stabilisation, water harvesting etc.) will also be important functions, and they often produce the greatest change for the least effort. Any potential threats, such as fire or flooding should also receive high priority.

So what are the key functions in this design?

- *The most important things that your clients want.*
- *Those remedial strategies (points of intervention) that will plug the worst energy leaks (spirals of erosion) on site.*

While you're learning this, you might select just three or four key functions to focus upon. Then as you get more proficient, you can become more ambitious and choose more. That said, you'll soon notice that some functions are actually there to support others. So your key functions will be those client-led ones like food production and the associated ones those that protect the ecosystem in supporting them, such as soil conservation and improvement, wind protection and irrigation.

Here are some functions that may be relevant to your design:

Accommodation, shelter, privacy, warmth, children's play area, food production, soil conservation, soil improvement, leisure, energy production, windbreak, water supply, irrigation, access, drainage, waste water treatment, fuel supply, wildlife habitat, fodder, education, community integration, cash income etc.

Aesthetics, while not originally considered in permaculture to be a *function*, might still be a factor which guides your choices. Harmoniously patterned systems blend into their environment and elements themselves also vary in aesthetic appeal.

Creating your SMART goals

OK, let's now take these key functions and turn them if we can into some SMART goals, by which we can judge our level of success at the end. Clarifying some details during the interview process will enable us to define our goals as:

* *Specific – being well defined and clear to everyone involved.*
* *Measurable – so we know when they have been achieved.*
* *Agreed upon – by everyone involved.*
* *Realistic – attainable, with given time and resources.*
* *Time-bound – with achievable but not too easy deadlines.*

By way of an example, the client might ask for an increased level of food production from their garden in order to be less reliant upon vulnerable shop supply lines. In this case we need to find out how much more food is wanted. Do they just want a few more potatoes every week, or full-scale self-sufficiency?

How might we measure that? In terms of the weight of produce, or the money saved on their food shopping bills? The way they define this could affect your strategies. The former might encourage the inclusion of a big area of mono-cultured potatoes and squashes, whereas the latter might instead encourage the creative inclusion of higher-value crops.

So our client's desire for more food from the garden might look something like this as a SMART goal:

> ***By the end of July 2010, Blossom cottage garden will be producing at least 200kg of food (for the client) every year.***
>
> *OR*
>
> ***By the end of July 2010, food grown in Blossom cottage garden will have reduced (the client's) food bills by at least 25%.***

It's as simple as that, but once we've done this for each key function, we'll have ourselves a clear framework to ulti-mately judge our degree of success by. So having identified the key functions we're going to design for, and created our SMART goals, we'll now think about the best ways to achieve them.

Choosing systems and elements

Ideally, we'll only be including systems and elements in our design if they fulfil at least three functions. Remember the ecological principle: ***Multiple functions for each element***. Nature happens to be so productive, because everything does so many things (take a moment to think of all the services trees provide). So we might design a road to also be a firebreak and a means of harvesting and directing rainwater into a dam. This saves both space on the ground and the additional cost of extra earthworks. It's also important to ensure that there's a diversity of elements in your design, providing security – especially for important functions like food, water, energy etc. Remember: ***Multiple elements for each important function***. If all our energy is supplied from one source (e.g. mains electricity) and that provision fails, then we're in trouble. If we don't have multiple sources to meet all our important needs, then we're very vulnerable.

The process

Instead of a 'black box' design, where we keep our workings in our head, we're going to follow a 'documented' route. Writing it all down makes it easier to see later why we've made particular decisions and is more likely to convince a client too. So, taking each of the chosen functions in turn, write down all the ways/things (systems/elements) you can think of to (realistically) achieve them. Disregard anything that's clearly way beyond budget, or too large or small scale for the purpose.

When you're done you should have three or four lists of systems and/or elements, one for each function. I often find that some of the things I think of are on several of my lists, which is a good starting point. I then go through my lists again to see which of the other systems and elements I thought of also perform other useful functions. This identifies some strong possibilities, though I still have to make sure that they're suited to the site conditions and other client requirements (e.g. do they fit in with the client's values?) before deciding to include them in any design.

Here are some criteria that I find useful in choosing the best systems and elements to include in a design;

Permaculture ethics

Perhaps the most important of all; how do the systems and elements we are considering perform against the three ethics?

> ***Care of the Earth*** – *rebuilding natural capital – soil, forests, clean air and water, species diversity etc.*
>
> ***Care of people*** – *physical, emotional, spiritual – yourself, your family, community, culture and humanity as a whole.*
>
> ***Fair shares*** – *of surpluses and voluntary limits to consumption and reproduction – equality leading to peaceful planetary co-habitation. Creating abundance within a finite environment.*

Ecological impact

Some of our options may have a much greater impact upon the environment than others. Factors that we might take into account to determine this are:

* ***Emergy*** (Embodied energy) – *this is the energy used up during the entire lifecycle*† *of a product. We don't need to know these figures exactly to make our choices.*
* *Biodegradability – natural resources are excellent choices for our designs, as they return to nature after use.*
* *Transportation – locally available resources use less energy to be moved to the site (weight for weight).*
* *Greenhouse gas (e.g. carbon) emissions.*
* *Footprint – how much land is required to grow something, though sometimes this might be measured as the volume of water used etc.*
* *Company ethics – much of the latest green technology is made by the biggest, dirtiest corporations.*

Refer to your PASTE list and make the most of the resources you've already found to be present on site or locally available.

† *This lifecycle includes raw material extraction, transport, manufacture, assembly, installation, disassembly, deconstruction and/or decomposition. Figures for commonly used materials can be found easily online – if you can't find any, then make a good guess. Thinking about the whole process is a good exercise in seeing the bigger picture.* The Story of Stuff *illustrates this well: www.storyofstuff.com*

The sectors you recorded are all about energy passing through the site from source to sink. Your job is to devise strategies to store as much of it as possible on its way through. This is what David Holmgren encourages us to do with his principle ***catch and store energy***. There's a whole lot of ways that we can store energy on site – naturally.

There's a guideline that reminds us we should:

* *First **obtain yields** by collecting some of that energy, perhaps in the form of plants or trees capturing sunlight, water held in the landscape, or wind made to do work.*
* *Next, take measures to **conserve** the energy and resources we have, by plugging any leaks and applying timely maintenance.*
* *Only after we've done these things should we resort to **using** up resources to replace what we've lost.*

Let's consider a home heating scenario. We can, for instance, harvest the sun's heat using passive solar gain and store it in thermal mass. We can insulate the building and incorporate measures like heat exchangers to ensure good ventilation without losing much of the heat with stale air Only where all these strategies fall short should we consider burning wood or fossil fuels to generate more warmth. Sadly most of us still have to, because we've inherited badly designed buildings that face the road more often than they face the sun.

Brighton Earthship

"If you feel like you need to leave, you're the person that needs to stay, because you're feeling things that are out of balance... you're the one who'll be the change-maker."[13]

Warren Brush

This means that you and I, having noticed that we're living in those badly designed buildings, are the best people to do something about them. After all, if we move out into a perfectly designed new eco-house, are the people that move into our old home going to eco-renovate it?

Look for opportunities to harvest energy and resources, and identify measures that will conserve them, minimising the need for inputs of extra work and energy.

Remember that ***the yield is only limited by the imagination of the designer***. Don't be afraid to seek out the ideas of others as part of your process. A fresh pair of eyes can see opportunities that, however experienced we are, we sometimes miss.

Now let's consider the ***Hierarchy of Resource Use***:[14]

* *Resources that **increase by modest use** – e.g. coppiced trees, grazed shrubs and grasses, even friendships (it's always possible to outstay a welcome!).*
* *Resources that are **unaffected by use** – e.g. good climate, stone as mulch or heat store, a beautiful view.*
* *Resources that **temporarily disappear or degrade** if not used – e.g. annual harvests including ripe fruit and nuts, water run off during rain, swarms of bees.*
* *Resources that are **reduced in the long-term** by use – e.g. mature forests, clay deposits, fish populations, coal, oil.*
* *Resources that **pollute or destroy other resources** if used – e.g. residual poisons, radio-actives, large areas of concrete, sewers discharging into watercourses or the sea.*

Try to only use resources from the first three categories. Any chosen from the last two should be done with good reason. In a similar vein we have the ***Spiral of intervention***[15] (sometimes also referred to as a *cascade*), which gives us a hierarchy of options starting with the least harmful strategies. Permaculture guides us to ***work with nature***;[16] so the simplest intervention is to ***do nothing*** and let nature redress any imbalance. Sometimes this involves the ***removal of limiting factors*** that are preventing natural succession, such as the fencing out of grazing animals (either wild or domesticated) or stopping the use of machinery or chemicals. Remember the Yellowstone wolves (p9)? They were nature's control on overgrazing of young trees by elk.

Should there be a need to ***accelerate succession***, the next safest level of intervention we can make is ***biological***. This means enlisting the help of plants, insects, birds, animals etc. as part of an Integrated Pest Management strategy.

Should these techniques be inadequate, we might apply some ***minerals*** or use ***mechanical*** means, which nature will still be able to repair, though over a longer period of time. Only when we've exhausted all options at that level should we consider the least safe option: ***chemical*** intervention. In theory, a skilled designer should never need to resort to the latter.

This intervention strategy is well suited to designing by ***Options and decisions***,[17] where the process can evolve in response to choices made about observations and the results of trials. We might use the cascade of intervention as our guide in dealing with a loss of yield or in regenerating degraded landscapes.

If, for instance, you were losing lettuces from your garden to slugs or snails, you might simply accept the loss (do nothing), perhaps appreciating how amazing they are and their place in the whole ecosystem or simply plant extra lettuces to compensate. If this were ineffective you might then introduce some decoy plants, or predators, perhaps chickens (biological interventions). You could apply rock dust to help the lettuces grow bigger (a mineral intervention). If these fail you might introduce physical barriers, moats, or copper strips (mechanical interventions). Poisoning the slugs and snails with pellets (a chemical intervention) should be a last resort because of its impact on the soil food web. By following this method, and not advancing from one step to the next unless it's completely necessary, we minimise our impact on the natural ecosystem.

When growing plants we might also consider some or all of Masanobu Fukuoka's ***Four principles of natural farming***:[18]

- ***No cultivation*** *– no ploughing or turning the soil. The earth cultivates itself naturally.*
- ***No chemical fertilizer or prepared compost*** *– these practices drain the soil of its essential nutrients.*
- ***No weeding by tillage or herbicides*** *– weeds are an important part of building soil fertility and in balancing the biological community. As a fundamental principle weeds should be controlled, not eliminated.*
- ***No dependence on chemicals*** *– weak plants develop as a result of ploughing and fertilising, increasing their vulnerability to disease and insects.*

Financial costs

Locally sourced materials were once the easiest and cheapest to use, which is why older buildings tend to blend into their environment so well. Sadly the use of cheap overseas labour and fossil fuel-driven mass production means that's not always the case now. Earth-friendly choices are often more financially expensive, locally grown organic food being a good example. We do however also have the restraint of a financial budget to consider and our task is to create a design within it.

Given that nothing is certain, I'd consider alternative options in the possible event of the client's available funds being either significantly increased or reduced at some point in the future. If you can offer 'plan B's to the client, it would add flexibility to your design. Also, while budgeting, consider how costs can be reduced by timely intervention (plugging leaks – '*a stitch in time saves nine*') or by buying cheaper options (e.g. seeds as opposed to plants). Indeed your whole implementation plan could be affected by strategies used to reduce costs.

Being able to cost a project is a really valuable skill, especially if you have any intention of designing professionally for other people. Don't forget though that for some clients *time* is their greatest limiting factor.

Time limitations

You should rule out any systems that will take too long to put in place, though a renegotiation of the project deadline might be worthwhile if an idea is a particularly good one.

Ultimately, being able to estimate how much time and money each phase of a design may require, will stand you in good stead. Getting into the habit of recording both the financial and time costs of any project you do, will make it easier for you to give estimates to future clients. If for instance, you know the time and materials used in making a raised bed of a given size, you can multiply those figures to obtain a reasonably accurate 'guesstimate' for other similar situations. As with all these things, practice makes perfect.

Appropriate scale

Consider what systems and elements would be best suited to the size of any site. A treebog would probably be too big for most urban gardens, while forest garden techniques could still be applied there using fan, cordon or espaliered trees. Ponder too the required capacity of systems. Can I grow enough of a particular crop in this space? How many animals will this land feed? How much power do I need and will this wind turbine / solar array be able to provide it?

Interconnectedness

By identifying the ***needs*** (inputs) and ***yields*** (products/outputs) of each system or element, we can identify where beneficial relationships can be created. If we can connect two elements together so that the outputs of one feed the needs of another, we are well on the way to creating a self-sustaining system. The more we can do this, the less work will be required to maintain the overall system we design.

Needs and yields (input/output) analysis

The classic textbook example is the 'Permaculture chicken',[19] where its needs, products and (intrinsic) behaviours are all considered. The ideal is to meet all its needs on site, make use of its behaviours to reduce work and also to make use of its additional outputs (e.g. manure), so avoiding pollution.

In order to identify possible beneficial connections, we need to do this for each key element in the system. As a result, this is a process that can take some time, but the time and energy that can be saved by good design is far greater. Our ultimate aim with this is to connect each of our best choice elements into a mutually supporting cycle. We'll return to this in the section on placement and integration, but for now we can use it to help us choose the most appropriate elements (e.g. plants, animals, structures, tools etc.) for our system.

Two other useful tools to help you identify elements most likely to integrate well with other parts of your design are '***random assembly***' and the '***web of connections***'. What we learn from these can also be of use to us when deciding about placements in the next section.

Web of connections

Draw a large circle and write the names of each of your potential elements around the outside. Choosing each in turn, go around the circle and identify where you can make connections with the other potential elements. As you find them, draw a line joining them together. In this simple example below, I've used a selection of colours to highlight the different kinds of connections; e.g. daily, weekly, seasonal, need to be supervised etc. You might even include another colour for 'elements to keep apart'!

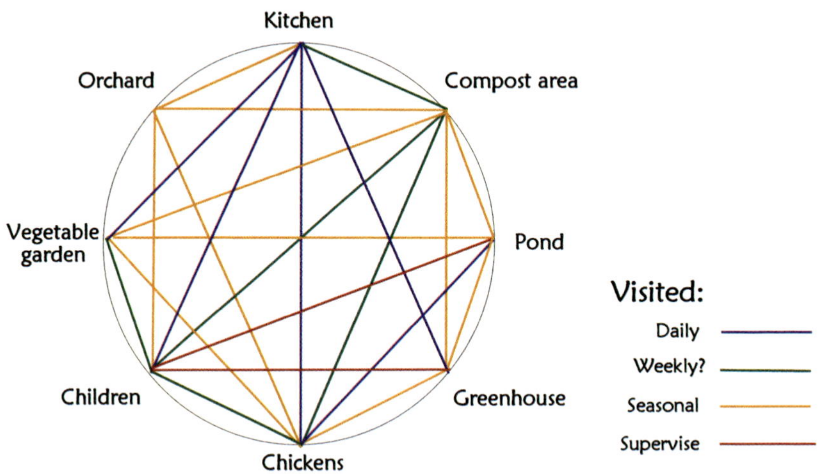

This example has only eight elements around the outside so it's easy to see; yours will probably have a lot more. In such a web, the most-connected elements tend to stand out as having more lines than the others. However, the most value is gained from going through the process, rather than looking at the final diagram. If you notice any really good connections, make a special note to remind yourself later, or perhaps draw a thicker line on the diagram.

How the elements connect will of course vary, depending upon how each system is intended to function. Here the children are being involved, perhaps sometimes supervised, with the garden's activities. In other circumstances, they might need to be kept completely away from the greenhouse or the pond. The pond in turn might take grey water from the kitchen indirectly via a system of cleaning wetland cells.

One other thing to remember here is that some of these elements are mobile and that others are fixed in place. Chickens are very mobile, a chicken shed probably won't be; unless you determine that it would be an advantage.† In order to make effective connections between less mobile elements, they will need to be placed at least in close proximity to each other, if not physically connected together.

Random assembly

This is a design tool that I've seen used in several different forms since it was introduced in *Permaculture: A Designer's Manual*. Here's my favourite way of using it. Write down each of the elements (or systems) that you have on your shortlist onto separate pieces of paper or card. Stack them into a pile, turn it face down, shuffle them and divide into two piles. Turn over the top card from each pile and see if you can think of any connections between the two systems/elements. Don't hurry, as this process can help you to identify connections that a quick consideration might not identify.

Random assembly: element picture cards and a connections prompt sheet used to identify possible beneficial interactions

† *Like in Joel Salatin's system where his flock follows a few days behind his cattle, feeding on fly larvae in their manure and spreading it about as they search for them.*

While many combinations won't elicit useful connections, this process is very good at occasionally inspiring innovative solutions to problems, the kind that others will remember as 'great ideas'. A recent example of this was when 'nut trees' came up with 'pond', a little consideration brought up the idea of planting the trees on an island, to help stop squirrels reaching the nuts. They can swim, but don't like being at ground level where they are vulnerable to predators. Again, you should make a note of any particularly good connections, or any elements that you think you need to keep well apart!

An addition you can make is to include a sheet of connecting words that say things like 'on top of', 'alongside', 'in the middle of' etc.[20] These can be useful prompts for thinking about the kinds of connections that could be made between things. Don't hurry this process, the best ideas can come from that moment when the conscious mind goes blank and inspiration strikes. Make a note of any good ideas you have before moving on to the next two cards. Now, if you've actually done this exercise, don't throw those cards away, we'll be using them again a bit later to help us with our placements.

Intrinsics

We started out by brainstorming a list of elements that could fulfil each function, but we may have not yet considered the intrinsic capability of each one to perform its function on this particular site. The efficiency of systems can be affected by climatic conditions and the availability of energy, resources and services, so we need to choose those most locally suited. Life in general has evolved around daily and seasonal rhythms, slowing down when supplies of energy are less available.

Human technologies are rarely different. Photovoltaic panels for instance, make electricity most of the year but only during daylight and in a DC form. Given a reliable water supply, hydro generators run 24 hours a day during wet seasons (the colder months in Britain, when demand for power is highest) and produce it in the more useful AC form. Choose technologies best suited to local conditions; their efficiency is irrelevant if the energy or resources they need are not readily available.

Likewise with plants, choose those that grow and yield well in the local climate. Do they prefer this kind of soil? Use your site observations, including jar and pH tests, shade maps and what you know about the growing season to make your choices. A few borderline tender species may be able to be grown in favourable microclimates, but consider them expendable luxuries.

Sometimes an initial investment can make life a lot easier in the longer term; for instance, making compost for use in an intensive vegetable garden. If we can grow more of our own food, we save having to earn so much to buy it further afield. In general though, choose elements that need less attention; they're often the best choice, even if they're less productive. Finally, when choosing multi-functional elements, pick those best at performing the function you most value. For instance, most ducks lay eggs and eat slugs, but some breeds are better layers and others are more voracious eaters of slugs.

Comparing best options

Hopefully we're starting to whittle our list down now to the strongest possibilities. Some choices are easy, others a little more difficult. Remember that ideally we want at least three ways of meeting each important function, so if we have difficultly making a choice between several strong contenders, our best option might be to include them all.

There will be other times when a choice has to be made, perhaps because of lack of space. In which case we can use a couple of simple thinking tools to help us: ***SWOC*** (which is adapted from the more familiar SWOT) and ***PMI***. Both these are simply frameworks to help us make comparisons. Having applied either of these tools to our different options, we should end up with a shortlist of those offering the most potential.

SWOC stands for:

> ***Strengths*** – *what are the good things about choosing this?*
>
> ***Weaknesses*** – *what are the not so good things?*
>
> ***Opportunities*** – *what will this also allow me to do?*
>
> ***Constraints*** – *what negative effects will this choice place on the things around it?*

For instance, we might decide that for a windbreak, a hedge would offer us a solution that gets more effective over time (S), but takes a while to establish (W), it could offer additional outputs like wildlife habitats and food/fodder (O), but create shade and competition for other plants growing on the shady side of it (C). A wooden fence might provide a fairly instant barrier (S), but need more maintaining (W), provide a good vertical structure for climbing plants (O), but involve damaging those climbers when maintenance takes place and also throw shade to one side of it (C).

Basically, the SW is about the thing itself & the OC the effect it has upon the things around it. Call it COWS if that makes it easier for you to remember!

Or you could use a simpler version of this, another of Edward de Bono's thinking tools ~ ***PMI***, which stands for:

> ***Pluses*** – *what are the good things about this?*
>
> ***Minuses*** – *what are the not so good things?*
>
> ***Interesting*** *things – what else might be relevant, even though they may be neither good nor bad?*

SWOC gets you to consider the relationships between elements, whereas PMI lacks this emphasis, but it doesn't matter how you decide which of them to choose, only that the method gives good results. Use whatever works best for you.

Visioning and wild design

Our approach so far has been very analytical (it's what this section is called after all), but don't forget to leave time for your imagination. This requires a totally different environment from the rest of this section. Random assembly has allowed your mind to make creative connections, but only between the elements that your logical mind came up with. Hopefully you gave your clients the chance to express their own vision, now give free rein to your imagination. With your logical brain now satiated, permit yourself time, ideally back on site, to walk around or just sit and contemplate. Allow your mind to just wander and let nature speak to you. Let the holistic processor in your head take over for a while and see what emerges.

Summary

The key things to remember about the analysis stage are:

Analysis

- *The analysis phase is important and should receive as much time and attention as the observation phase did.*
- *How you make decisions is up to you, but good design is underpinned by the use of effective, proven processes.*

Identifying functions

- *Functions are what you are trying to achieve, systems and elements are how you will achieve them.*
- *Your client interview (e.g. their needs and wants) and site survey (e.g. limiting factors) will identify the key functions of your design process for you.*
- *Some functions will be primary ones, supplying the client's needs directly (e.g. increased food production); others will be secondary, providing support (e.g. soil building, erosion control, irrigation) for the former.*
- *Some may be required to protect from potential threats.*
- *Try to turn your key functions into SMART goals.*

Choosing systems and elements

- Follow a documented route to remind yourself later of your reasoning and to gain the client's confidence.
- Get all your ideas for fulfilling the functions, out of your head and down onto paper first. Then use the following processes to pare them back to your best options.
- Which of your possible elements are multi-functional?
- The ethics should be the next filter for your initial ideas.
- Consider the footprint/Emergy etc. of each option.
- Identify the biological/renewable resources on your list.
- Review your PASTE sheet and consider how many on-site or local resources could be made use of.
- Remember: Collect, Conserve and only then, Use up.

* Review sectors and look for opportunities to harvest wild energies across the site.
* Consider where your different elements lie within the Hierarchy of Resource Use and the Spiral of Intervention.
* Ensure that systems are of an appropriate scale for the site.
* Will all the remaining options keep you within budget?
* Can they realistically be achieved in time?
* Decide which of the elements are the best connected, using tools like Input/Output analysis, the Web of Connections or Random Assembly.
* Make sure that your choices are best suited to perform the required functions on this specific site.
* Compare your shortlist using tools like SWOC or PNI.
* Are you meeting each important function in at least three different ways?
* Give your imagination free rein to explore the abundance of possibilities that exist for the site.

A multi-functional salad: living mulch, food, celebration, art and income source

Choosing systems and elements flowchart

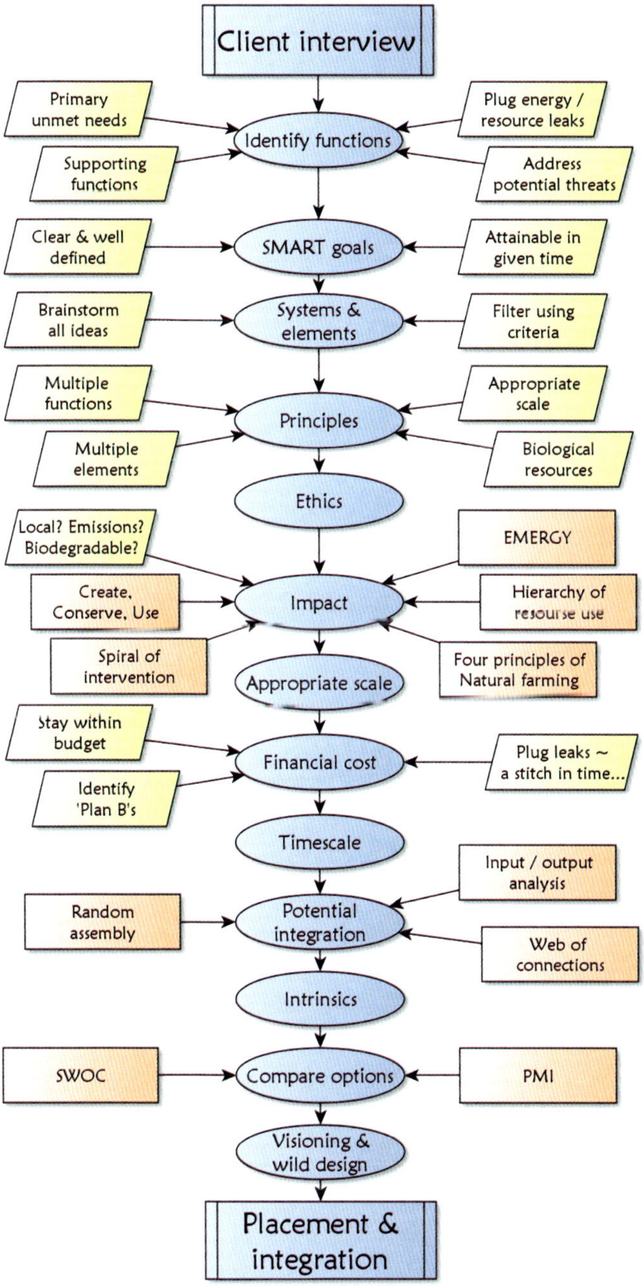

Placement

Now we'll experiment with best placements for the different elements and systems in our design. If there's a fixed point of focus on the site (such as a house), then we'll be placing everything most efficiently in relation to that. However, when starting with a 'blank canvas' we get to choose the best place to site our centre of activity. If we're arranging our design around a proposed new house or other structure, its placement will probably be our most important decision. If we're designing a site without an obvious central element, then we'll need to identify the most important elements to place first and go from there.

Around an existing main element

When designing around a fixed point of focus like a building, we've a number of methods we can use to plan the layout of the site. Using our base map and overlays from the survey, we'll aim to make our mistakes 'on paper', instead of in the landscape itself. One helpful tool for this process is the ***land-scape modelling*** technique I use to teach zones and sectors.

Using landscape modelling to explore placement of systems and elements

Laying out an approximate model of the site on a floor or table using cushions, cloth and ropes, gives us a 3D space into which we can explore placement of our systems and elements. We can use the same cards we made earlier for 'random assembly', to represent elements, and by moving them about, explore the relationships between them. Creating a model can give us a landscape overview that we don't get from ground level. Imagine you're 5,000 feet tall or flying over the landscape in a hot-air balloon. Now, consider the following factors in turn to identify the best location for each system or element.

Design from patterns to details

Before looking at the finer details of your design, establish an overall pattern. This will be guided by many of the factors that follow. It's important to adapt your design to the opportunities the site offers, rather than trying to adapt the site to the design (a common human strategy now we use fossil fuels to replace skill with brute force!). You may be excited about creating a forest garden, or growing olives, but on an inappropriate site you'll be wasting time and energy fighting nature. For instance, hilly and mountainous regions are excellent for storing water high in the landscape, where it can then be made to do work (e.g. hydro-electric power generation or irrigation), but not so good for growing tender crops. Identify the opportunities the site offers and design to take best advantage of them.

Microclimates

This may seem like jumping straight ahead to the small stuff, but microclimates actually come in all shapes and sizes. The whole of Britain is a microclimate, benefiting as it does from the warm ocean currents delivered by the North Atlantic drift. South-westerly prevailing winds bring plenty of rain and the west of Britain receives more of it than the east. In turn, Dartmoor, the area of high moorland in the south-west region, has even more rainfall than its surroundings. Focusing in further, we can pick any of the valleys on the moor and find sunnier and shadier aspects, windier and sheltered spots, damper and drier areas. Look closer and you'll continue to see variations, until you pick up a single stone and find creatures there that like the dark and damp conditions it provides. Our job as designers is to spot these opportunities and place things there that will make the best use of the gifts they offer.

In Britain, gardeners have long taken advantage of favourable microclimates offered by south-facing walls to grow tender but highly valued crops like figs and grapes. Here's another ingenious version of this strategy but with additional functionality, devised by a great friend, Tony Martin. He not only uses the thermal mass of his garden water storage tanks to protect the crops on his northwest-facing hilly site, he also uses them as a reservoir for a capillary watering system. By growing vulnerable but high value crops like strawberries on top of the tanks he keeps them away from hungry bugs. The plants and soil also protect the tanks from much of the sun's degrading UV radiation and reduce algal growth inside. Genius! Full details of how to make them are in the online resources.¤

Perhaps the greatest skill lies in making best use of the colder, shadier and damper areas. Those are places where most life slows down, though we can still use this to our advantage. Those cooler spots such as shade-facing walls, are ideal for the things we want to keep cool to preserve them, such as food stores. Shady spots have more stable temperatures, so young plants and cuttings are less vulnerable there in winter months than in direct sunlight. Because shade-facing slopes receive less sun, trees take longer to wake up in the spring. The resultant delayed flowering can make all the difference to any early blossoming trees such as pears and cherries, when late frosts could damage flowers and ruin a fruit crop. In the tropics, shade is often sought out because photosynthesis shuts down when it gets too hot and the evaporation of water is reduced where there is less direct sunlight. We discover where and when all these microclimates occur by ***protracted observation***. This is where our diary reveals its treasures, especially the notes we made during any inclement weather. Shade maps and photographs can also help us to identify the right microclimates for the elements we are placing.

Elevation planning and aspect

By considering the altitude of a site and any slopes present, we can identify the diverse microclimates, opportunities and limiting factors that occur there. There are specific strategies that work best at each point on a slope, such as using trees to stabilise steep gradients and having vulnerable food crops growing above the frost line. The thermal belt between frosty hilltops and valley bottoms in cool temperate zones offers a slightly longer growing season. The aspect of a slope is also a factor to consider. For instance, growing conditions near the top of a sun-facing slope can equate roughly to those at the bottom on the shady side of the same hill. In other words, the extra sun compensates for the limitations of the higher altitude (temperature drops about 0.5°C for every 100 metres above sea level).

So how do the design elements need to be arranged in relation to slope? If we want anything to flow under gravity through a system, we'll need of course to place them at the correct relative heights. To move water from roof guttering, into a raised tank, then a toilet cistern, on to a septic tank and finally a wetland treatment system, each needs to be placed below the previous. Water can be pumped uphill using technologies such as the hydraulic ram, but it's simpler to work with gravity as much as we can.

An excellent solar hot water system at Ourganics making full use of both gravity and the thermosyphon effect

Similarly, a solar hot water panel is better placed below a hot water storage tank in order to take full advantage of the thermosyphon effect and removing the need for a pump. Warm air rises, so in buildings we should consider this in room placements and heating systems. It uses far less energy to draw warm air down from a ceiling than to heat the cooler air at floor level where we sit down to relax.

Zoning

Systems and elements that need the most attention are of course best placed close to our central focus, where we spend most of our time. Start by identifying which elements in your design will need most attention. Do this by thinking about how often a particular thing needs to be visited, either to clean, repair or harvest from it. A chicken shed for instance should need at least two visits a day: to let out and shut in the chickens. In addition there may be separate daily visits to collect eggs. Cleaning occurs perhaps a couple of times a week. Shed maintenance may be a few times a year. By way of contrast, a compost heap might have fresh green waste deliveries from the kitchen only every few days, need turning about once every ten days (for hot systems) and be visited a few times in the spring to harvest the compost. From this we begin to notice that some tasks take place on a daily basis and others may be more seasonal. Use this information to place elements needing most attention closest to your central point. This is a good general rule to follow, though there are exceptions, as we'll discover in the next section.

Growing my salads in zone one

Access and desire lines

Access points and desire lines will influence the basic circular zoning pattern; regularly walked routes can also be considered as zone 1, allowing any higher maintenance systems to be sited there. Driveways are given less attention as a driver's mind is often on where they're going, or what they'll need to do on arrival at home, so design them as low maintenance areas. Conversely, you might deliberately place high maintenance systems or points of interest where they create new advantageous desire lines. An attractive new feature in a previously neglected area will get more attention, making extra zone 1 space along the path leading there.

Don't make that path too straight though if you want people to slow down and interact with elements along the way. We obtain further benefits by creating new desire lines for other creatures too. Place a high pole in the middle of an open area where you're growing vegetables to encourage birds of prey. By giving them somewhere to perch, they can help to manage rodent populations for you as another part of your Integrated Pest Management strategy. Sometimes the best placement of elements can make life so much easier. Consider how much less work it is to herd animals if gates are placed in the corners of fields instead of half way along a side.

Sectors

Now let's consider how incoming wild energies, such as wind, water and sunlight, might interact with each of our chosen systems and elements, which we can position to either harvest those energies most effectively, or to shelter other elements from them. Often, the best placements for creating a yield are where wild energies are already being focused, such as harvesting water by building a dam where an area of sloping land directs run off towards a Keypoint (arrowed). Notice how much larger the trees are, both there and in the valley below, as the water and the fertility it carries starts to settle out on the gentler slopes. As it takes a lot of energy to change the landscape (remember the *Scale of Permanence*), it's far easier to work with the existing terrain. This is why Keyline® lays out the water systems first and irrigates the land using gravity, through the use of subsoil ploughing and flood irrigation along gentle downhill gradients from valleys to lower elevation ridges. These techniques are well suited to obtain a yield from the opportunities the topography of the land offers.

A keypoint on a Greek hillside

In a similar way, upslope wind is compressed and accelerated towards the brow of hills by the rising ground below. We can use this to our advantage by placing wind turbines there. This is an example of harvesting the energy of a sector. Of course, when considering most buildings, we'd instead be looking to provide shelter from those winds to reduce heat loss. In the higher latitudes, it's also important not to block low winter sunlight from coming into buildings, so aim to keep that sector clear of anything that would reduce it. Compromises often have to be made though, such as between allowing all winter sunlight to reach a building and completely protecting it from prevailing winds. In Britain the latter come from the south-west, which is also the direction of the midwinter sunset. Which you choose to favour will probably depend on additional factors. Certainly for plants and trees, wind is the main limiting factor to growth and for pollination by insects, which don't fly when it's windy. Thus it can be a good trade off to sacrifice a little sunlight for a good windbreak.

A similar issue arises where a good connection exists between two elements that need to be physically close together, but prefer different microclimates. Low voltage wind turbines are often used to provide power to buildings, though the windy conditions that suit energy generation can reduce the energy efficiency of any building. Placing them further apart in order to overcome this increases the distance the electricity has to be transmitted, which for low voltages can result in high losses. This leads us instead to considering larger, higher voltage wind turbines more suited to serving a collection of dwellings.

To give an example here of the importance of good placement: some orchards use energy-hungry giant fans to stop cold air settling on blossoms and ruining the crop. Nature wouldn't put those trees in that frosty microclimate and if we want to avoid a whole heap of wasted energy, neither should we. Lastly, consider how to protect and enhance any highly valued sectors on the site like good views and places of tranquillity and how those unwanted sectors like bad views, noise and pollution can be moderated by good placements. Sometimes a system or element's ideal spot can be determined by simply noticing its optimum zone, sector and elevation.

Soil type

That said, when considering plants and trees, growing them in the right soil type is also important, so this is where we review what we learned from our observations. If soil conditions vary across the site, then we've a greater diversity of opportunities for growing. As gardeners we're often told that an ideal soil is a good fertile loam with a pH around 6.5, but if the soil was like that everywhere we'd see far fewer species around us. The Earth's varying soil conditions lead to a diversity of ecosystems containing a multitude of species and that extra diversity leads to more beneficial relationships, making the whole web of life stronger. ***Diversity creates stability***.

Now, the evaluation of zones and sectors may suggest growing a particular plant in a given place, but if the soil is of insufficient fertility, the wrong pH, or too wet or dry, we should probably think again. It's always easier to find plants best suited to the local soil (as nature does), than to try and modify it. That said, sometimes (with extremely acid soils in particular) it may be worth the investment of time and resources† to modify pH, at least in a small area, as part of a general soil improvement strategy for intensive food growing. This is almost always the case for zone 1.

Utilities

Don't forget that any elements needing mains utilities (e.g. electricity, gas, water, sewage, telephone etc.) will need placing where they can be easily connected, unless of course the benefit of choosing a particular site outweighs the cost of extending cables or pipes.

Visualise succession

> *"Vision is not seeing things as they are, but how they will be."*
>
> David Holmgren

Nature never stands still. Trees grow and throw more shade, microclimates emerge and habitats change. Old trees fall down, rivers change course and then there's climate change to consider. Depending upon the timescale of our design, these may be things that we need to consider when deciding placements.

† *Particularly if those resources are unused by-products such as wood ash.*

Observing the landscape and ***visualising succession*** enables us to anticipate changing conditions and create a design that evolves to take advantage of them. This is a skill increasingly acquired with experience. Sometimes the client's vision includes distinct phases. For example, an initial focus may be on erecting a dwelling, but with need for an intensive food-producing garden at the same time. You might start by placing the garden safely out of the way of building work, and only move it to zone 1 once the building is complete. A forest garden could then be established where the intensive garden previously was, suggesting the planting of young fruit trees there from the beginning and establishing the intensive food garden around them. Visualising the development of a site gives us clues that can help us accelerate the succession of any systems we install.

Incremental design

"Allow the system to demonstrate its own evolution."

Bill Mollison

A design drawing only provides a snapshot of how a site might look at a specific stage in its evolution, so visualise how the site is likely to mature. In addition, the needs of the client may also change. They may develop an increased dependency upon the site for food or resources, either due to scarcity of them in shops or a greater number of mouths to feed.

Future proofing: a new orchard at Ragman's Lane Farm, planted to reduce their reliance on buying in apples for their successful juicing business

Perhaps surpluses from other local growers reduce their income from what were previously high-value crops, forcing a change of direction? New people may bring new ideas. The site may be required to fulfil extra or different functions. Neighbours may change and this may bring an increased risk of pollution from the surrounding environment. Some changes can be anticipated and some cannot. By involving the client(s) as much as possible during the design process, we ensure that they're better able to observe the evolution of the design over time and to respond accordingly. A design can never be completely finished, as it will get tweaked over time to improve its performance.

An example of this is the Welsh 3,000s race, a route of around 26 miles from the top of Snowdon to Foel Fras, including some 13,000ft of ascent and 14 summits over 3,000ft. Over the years, people have tried varying the route in different ways to shave a little time off the record. Each new successful variation improves the design, but as there are an infinite number of fine tunings that can be made, the perfect route may never be found. This is classic incremental design.

The fine details...

Once we've laid out the basic pattern for the whole site, we can start to look at the finer details of each system; this is where we create our planting plans and so on. It's useful now to take our ideas and start trying them out on the site to check they work in more than just theory. By laying out our plan on the ground, using ropes, hoses or stakes as markers, we can get a sense of how our proposal looks in the real world. Invite your clients to walk around the pathways, ideally for a week or so, and then give you feedback (this could be done by moving the markers). Confirm it all when it feels right to everyone.

The Blank Canvas

From time to time we find ourselves with the additional freedom (or challenge) of designing a 'blank canvas'. Without an existing central element around which to identify any zones, we'll give more consideration to those other factors to determine the best place or places around which to focus our design. I find the following technique useful when deciding where to place zone 0.

McHarg's exclusion method

This is a useful tool to help us place our most important elements, any fussy ones needing particular inputs, or those adversely affected by multiple factors. By systematically eliminating areas or sectors not best suited to our chosen element, we can quickly narrow down options and simplify identifying an ideal site. Sometimes there are so many 'no-go' areas that we're left with only one option.

This was the case when I sought to identify the best place to plant apple trees on an Irish mountainside; once I'd ruled out the areas that were too wet, too windy and potentially frosty, I'd only one area left. Of course you may be looking to place an element that's adversely affected only by frost, or waterlogged soil, or cold winds, or by some combination of these.

This is where our overlays are really useful in making decisions, as we can choose to place only those with the relevant information over our base map. If this initial process still leaves us with more than one choice, we can then consider how all our systems and elements could integrate together. This allows us to eliminate areas unsuited to any systems and elements that we wish to integrate with zone 0.

For example, if we would like gravity-fed spring water coming into our home, then we can also discount any area of the site above the level of the spring for the build.

Determining the best place to plant apple trees on an Irish mountainside

Placement flowchart

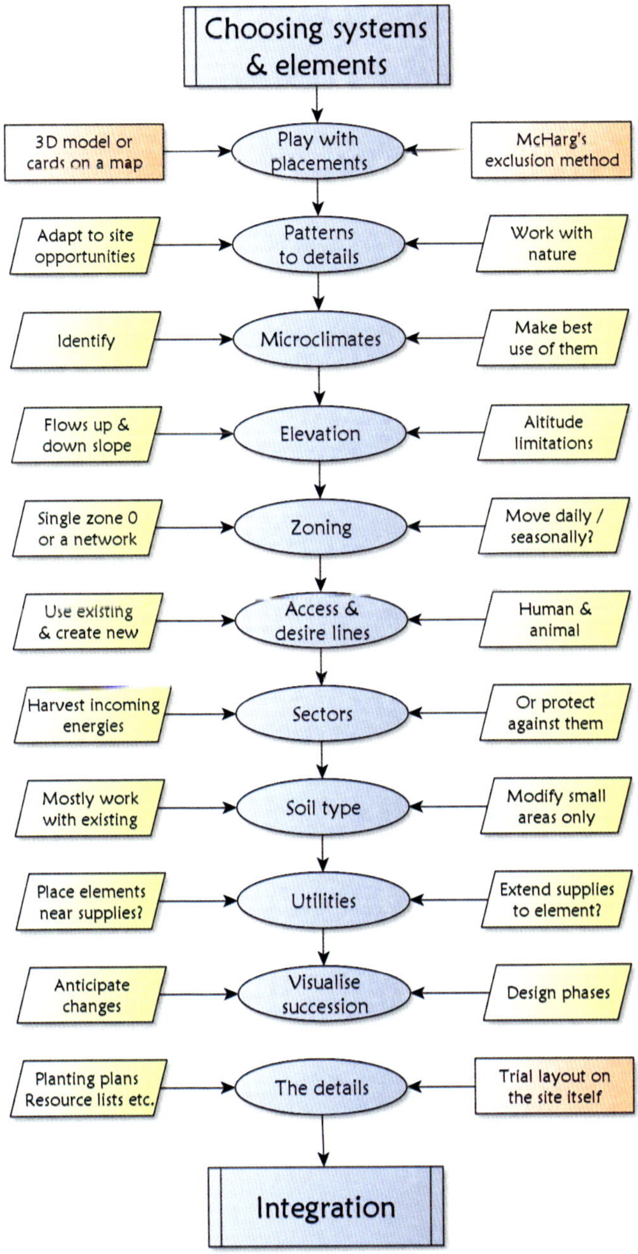

Integration

So having placed all our systems and elements where they each appear to work best, we'll now look at how they can be integrated together and if that suggests any better layouts.

Creating plenty of ***beneficial relationships*** will endow your design with resilience and efficiency, so look to connect your systems and elements together wherever you can. For living components (e.g. a chicken) it is often enough to simply put them in the right place; they then tend to create beneficial relationships with whatever is around them. With non-living components (e.g. solar hot water panels) we have to make the connections ourselves. As before, there are a number of tools that we can use to guide us in making these decisions.

Energy cycling

> *"Our work as permaculture designers is to prevent energy leaving before the basic needs of the whole system are satisfied, so that growth, reproduction, and maintenance continue in our living components."*[21]
>
> Bill Mollison

So design to make the best use of any energy or resources moving through a site. Link as many elements together as you can to create self-sustaining cyclic systems and to build fertility or produce a surplus, in the way that nature does so well. Any needs and yields analysis you did in the previous section will help you in this process. By connecting elements you can reuse a valuable resource like water several times before it eventually leaves the site (in a non-polluting manner of course). In the case of water, we need to identify which elements require the highest purity, so we can chain them together in the right order.

Then by mapping the connected elements onto the site we can attempt to move the water through the system using nothing more complicated than gravity. This may not always be possible for the whole system, but wherever you can use it, it is of course the most reliable technology for the job.

A Permaculture cup of tea

Above is a collection of cycles set up to apply permaculture design to our familiar cup of tea. These days tea drinking has become quite an industrial affair, with large amounts of energy and resources being used to produce, pack and transport the tea. We mostly boil water by using electricity, which in turn is often generated by the burning of fossil fuels or through dirty nuclear reactions. Then we visit the toilet and the vast majority of the time that fertility is lost to the site.

In the above example, the tea plants are grown on site; easily done outdoors if you live in a warm climate, but in Britain a tea bush would need a conservatory or greenhouse. Alternatively we could acknowledge that drinking infusions of *Camelia sinensis* is simply a habit and one that can be changed.

There are plenty of herbaceous plants native to cool temperate zones that make perfectly nice tasting infusions, albeit without that stimulating caffeine hit. These we can grow with minimal work and propagate easily, recycling used tea leaves and kitchen scraps to make compost and feed back to the plants. Diluting urine, which is high in nitrogen, makes a great fertiliser for leafy plants – just what we need! Water can be harvested from roofs, though it would need filtration before drinking.

As for heating water, well that's one thing that most types of tea need. Best options for us in cool temperate zones might be to boil the water on a wood burning stove or range (but only available during winter?), cooking using biogas, or by using a rocket stove or Kelly kettle, though the last two produce a lot of smoke and may not be ideal for indoor use.

Finally, it's also worth noting that creating cycles can help us by highlighting any important elements that we've overlooked, in the same way that doing a jigsaw brings a missing piece to our attention. This might also highlight new opportunities that have previously gone unnoticed.

Process flows

While the ideal is to create localised cycles to build on-site energy and resources, some will extend beyond the boundary of our design and involve exchange with the outside world. That's OK; life trades resources all the time and it's a perfectly normal thing for us to do, in fact it creates the resilience of nature's web. We just need to remember to trade equitably, both with our fellow humans and with life as a whole. So to make something of value to trade, we might bring in some off-site resources in addition to harvesting wild energies and utilising those we already have on-site. One example might be importing manure to enrich the soil and help us grow fruit and vegetables to sell. In such a case, try to arrange it so that each end of this chain has good access on and off the site. When the manure is delivered, you ideally want a drop off point close the garden or there'll be a lot of extra barrowing work to do. At the other end of the chain, position any sales area or collection points close to a good access too.

Where the other tasks in each process take place may be determined chiefly by their need for proximity to access points. However, one or more process tasks may need resources, tools, or energy available only at a specific place on site. Identify these tasks and aim to create an efficient chain through them from beginning to end. We'll be looking at how we can use flowcharts to plan such things in the section on designing beyond land-based systems, but remember for now that process flows can help determine placements.

Patterns and principles

> *"Patterning is the way we frame our designs, the template into which we fit the information, entities, and objects assembled from observation, map overlays, the analytic divination of connections, and the selection of specific materials and technologies. It is this patterning that permits our elements to flow and function in beneficial relationships. The pattern is design, and design is the subject of permaculture."*[22]
>
> Bill Mollison

As well as defining the overall design of the site, we can use patterns to help us integrate our elements into systems. This is where we consider which of the successful patterns that we see occurring in nature are best suited to our particular circumstances. Where, for instance, can we apply the principles of ecology, energy cycling being one such example? How can we create beneficial relationships? Where can we make best use of succession or stacking to integrate elements together?

Even when there's not space for all the layers of a forest garden, we can still make use of vertical and temporal opportunities. Can we create new microclimates and make best use of existing ones? The same applies to patterns of behaviour. How can we design our site to suit the habits of both humans and animals? Where are the desire lines (where they like to walk)? What tasks are routinely carried out together, and could they be integrated with each other around these lines? What's already working well together?

Martin Crawford's forest garden makes good use of stacking

The information that we find in gardening books and on seed packets about sowing times and soil preferences are simply successful patterns. If we had as much seed as say a tree produces then we could afford to throw it around everywhere and see which ones grew, but as we generally don't we can make use of these patterns to increase the chances of the few we have.

These are patterns that have worked well many times in other places, but ultimately you will learn about your site from first hand experience. Similarly, a ***Pattern Language***[23] has been developed to assist architects, landscape architects, planners (and us) to design more people-friendly buildings and townscapes. Biomimicry is an exciting example of the application of natural patterns in developing sustainable technologies. Biologists are now working with designers to show them where nature has already solved the problems that they are trying to address. The results so far have been very promising.

Edge effect

One such pattern that we see in nature is the edge effect. We know that the boundary between systems is a fertile place, because this is where interactions take place. Straight edges are rare in nature. Rivers take a winding course to deliver life-giving water and fertility over a wider area. Natural ponds and islands have extensive edge. Trees branch to increase their exchange with the sun and the atmosphere. All are more integrated and efficient because of this.

The edges we see in nature have a particular depth; the effect of interactions across them diminishes with distance. The depth of a woodland edge may be chiefly determined by the amount of light penetrating the canopy from the side. We also know there is a seasonal aspect to this with the spring flowering of many ground cover plants under the trees before the canopy comes fully into leaf. Spring is effectively the temporal edge on which these plants live.

So the depth of an edge is defined by the ability of resources or services (sometimes called vectors) to travel across it; bear this in mind when designing them. Quite how far apart two elements can be (across an edge) and still gain some benefit, depends on the means of interaction. How far plants can grow from a water source might depend on how extensive their root systems are. Benefits like pollination can occur over a much greater distance; insects might carry pollen as much as half a mile or more, the wind even further. Birds or squirrels might carry seeds or nuts out of the woodland into the meadow and distribute them some distance away.

Some edges can be significantly deep, though defining the exact distance where an effect becomes negligible is difficult, as it can vary with changing environmental conditions such as the weather or soil moisture levels. Just think how you might choose to cycle into town on a sunny day, yet take the car instead when it's wet and windy. Pollinating insects behave in a similar way! Such things are extremely fuzzy.[24] Beneficial effects can be short-term or year round, depending upon the means of interaction.

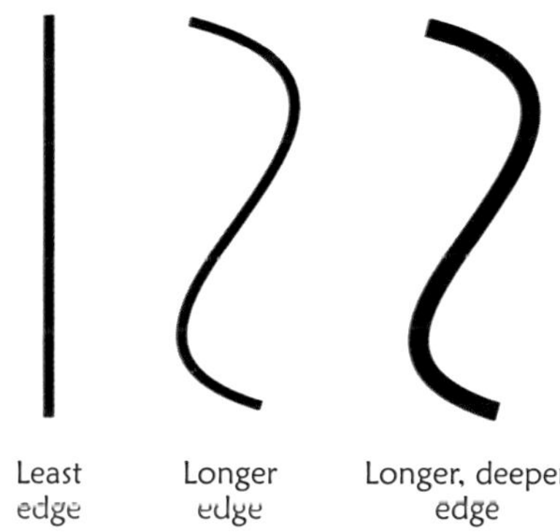

So when increasing edges we can do so by making them both longer and deeper. Nature's patterns, like crenellation (wavy-ness) already show us how to make them longer. Depth can be increased by wherever possible helping the agents of transmission, like planting a windbreak hedge for pollinating insects or by buying a good waterproof jacket for a hesitant cyclist! By creating ponds with gently sloping edges we increase the area of shallow water that warms in the sun: a popular micro-climate where creatures such as tadpoles thrive. Adding mycorrhizal fungi when planting most plants and trees creates a beneficial relationship that increases their ability to 'reach' soil nutrients.

Remember though that while increasing edges in our designs is generally a good idea, that we should always *optimise* rather than maximise them. If we produce more of a resource than we can make use of, we create more work for ourselves having to harvest it all and find new uses or outlets for it. If we leave it we risk creating an imbalance in the system, any unused resources having the potential to become pollution and attract pests. Don't be greedy, match your design to your needs. You can always increase that edge again later when you have the capacity to deal with the extra yields. Remember, ***start small and work out from well managed areas.***

...but don't be an obsessive *edgetarian*.

Strip farming

Strip farming is a simple technique for applying edge effect to growing crops, that requires little change from existing monoculture practices. Two (or more) crops are interplanted where there's a benefit from doing so. So how close together do species need to be (we might call this the effective distance of interaction) for a beneficial effect to occur? This is something we often find out by trial and error as it will vary depending on local conditions. Deciding this will allow us, for instance, to determine the ideal distance between pollinating trees in an orchard or the optimum width of strips of our poly-cultured plants. Strips should be no more than twice this effective distance across, so that even plants in the middle of each strip are close enough to the adjacent crop to receive a beneficial effect. Then again, strip width may be determined by nothing more creative than the width of the farmer's machinery.

One such example involves wheat being grown in strips with alfalfa, a perennial legume (nitrogen fixer). By growing them together, the wheat benefits from the nitrogen added to the soil by the alfalfa's annual root shedding. The alfalfa also acts as a windbreak to prevent wind erosion of the soil after harvesting of the wheat and the wheat does the same job for the alfalfa in the spring. Research has also shown that alfalfa attracts pest predators for wheat,[25] which is an additional yield. The beneficial effect is so pronounced that this particular system can apparently produce as much wheat grain as if the whole field had been used to grow it and produce an additional alfalfa fodder crop too.

Alley cropping

A similar pattern is utilised in designing agro-forestry alley cropping systems, where rows of trees are planted between crops such as grains and vines. Ideally, trees are planted at a lower density than usual to allow plenty of light in between. Crown lifting the trees can help too. While the simplest systems may involve growing just one crop with one species of tree, several crops could be grown together in this way. In addition to mining deep minerals and delivering fertility to the topsoil through leaf drop, some trees may also be nitrogen fixers for the system, others as fruit or nuts crops or long-term timber.

Strip widths may be optimised for most beneficial relationships between neighbours, though shade is also a consideration, as well as perhaps the width of the harvesting equipment again. In higher latitudes such as Britain, rows are often planted north-south to make best use of the midday sun, though lines of trees could instead be part of a Keyline®, swale or windbreak system, so consideration might need to be given to (interaction with) slope and sectors. It's also worth noting that the straight lines we usually see in such systems have less 'edge' than wavy ones.

Guilds and companion planting

Bill Mollison defines a guild as a:

> *"...harmonious assembly of species clustered around a central element (plant or animal). This assembly acts in relation to the element to assist its health, aid our work in management, or buffer environmental effects."*[26]

We see these assemblies everywhere in nature, because when seeds fall near those of other beneficial plants they will grow much more abundantly, those guilds ultimately dominating the space. Animals also thrive where the plant species that support them already grow.

Just because something works well in one place though, doesn't mean that it will do so elsewhere. While the *three sisters* guild (beans, corn and squash) works well in the tropics, it doesn't do so well in Britain's wetter cool temperate climate; the beans tend to grow faster than the corn. A number of companion planting books and charts are available that can provide you with ideas to try yourself, but their recommended combinations may have been quite specific to a particular site. Those authors may even have overlooked other important nearby species, without which those benefits would not have occurred.

Mint planted under an apple tree as a bee attractor

Remember, systems are complex and the boundaries we often put around them are arbitrary and illusory. Basically, other people's plant combining experiences can be useful starting points for discovering your own, but remember that your site and ecology will be in some way unique. When creating your own localised guilds, a good place to start is to place a nitrogen fixer (often a legume) together with a heavy nitrogen feeder and another species that helps control their pests. Play with different combinations and see what works for you. Pay attention to wildlife corridors too; sometimes the guilds you've created rely upon intermediaries from zone 5. Give wildlife safe routes through your site and ensure that any barriers you create take account of existing animal desire lines. This is especially advisable with badgers who will wreck any fence you erect across their regular runs!

Co-action analysis

Designing beneficial relationships in just two directions, such as strip farming or alley cropping is a relatively simple process. Doing the same with more complex systems like guilds and forest gardens requires more planning. Any plant or tree in the design will have other species around it in multiple directions. This makes planning a layout more complicated, but the same idea applies to placing together species that interact beneficially with each other and keeping those that are antagonistic apart. As we know from our quick look at systems, placement is everything. One tool that can help us with this is co-action analysis.[27] In our example, we have two species of tree that we wish to grow. The larger tree exudes a biochemical that suppresses the smaller tree's growth (this is indicated here by a '–'), but doesn't benefit in any way from its proximity (shown here by a '0').

Not a great arrangement, so far, but allelopathic† relationships like this are quite common in nature.

† *Allelopathy is a biological phenomenon by which an organism produces one or more biochemicals that influence the growth, survival, and reproduction of others.*

We could choose a third species to plant between the trees to buffer the allelopathic effect of the larger tree and perhaps also provide additional pest-predator attractant services to the smaller one. This is better, but it can still be improved upon.

Through the careful choice and placement of a fourth species we can enhance the situation even further. This might be a dynamic accumulator† selected to pump specific nutrients and improve the soil. This gives us a much better set up; the new plant benefiting those adjacent to it while also buffering the third species from the larger tree's allelopathic secretions.

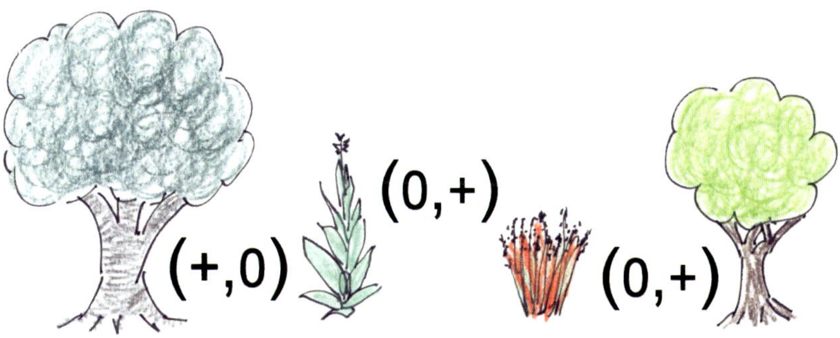

Now I've deliberately made this a hypothetical situation, as any example I might provide would be very site-specific.‡ Just use the concept to lay out species local to your own site.

† *Plants that are particularly well adapted, through their extensive root networks, at finding the specific nutrients in short supply in degraded soils. This gives them an advantage over other plants, meaning they often dominate in such conditions. Over time their presence delivers the missing nutrients through leaf fall to the topsoil, allowing other plants to succeed in the space. They're a good basis for liquid feed.*

‡ *In* Permaculture, a Designer's Manual, *Bill Mollison uses a guild of Walnut, Acacia, Mulberry and Apple to explain this placement process.*

If you have a list of species to include in your design, head out into your locality and look for places where any of them are already growing together. In each case notice how healthy each looks and then collate your results to identify patterns in their interrelationships. Such observations can be more useful to you than anything that you might read in books, which will have very likely been written by people living in different climates.

Biological control

Until now we've been looking to create mutually supportive guilds, but there may be times when you'll integrate species to deliberately suppress the growth of one of them. This is biological control and you might use it to naturally reduce pest or weed populations on a site. In its simplest form, you could introduce one species to feed on or out-compete another that you don't want. A stage on from that would be to introduce one species to attract another that feeds on the pests of a third. Of course, we're usually responsible for creating imbalances in the first place. The simple act of fencing a garden to keep out large grazers also excludes some predators. So we inadvertently create a sanctuary for the creatures they feed on – often those veggy lovers who like to munch on our greens. In such a safe haven their numbers will increase and the job of controlling them will fall to us – unless we get creative.

Forming plant guilds is fairly straightforward, however if we wish to include birds, animals or insects, we have to provide some incentives for them. For instance, to encourage bats to come and feed on the codling moths whose larvae eat your apple crop in the autumn, you might create a pond and put up bat boxes. The bats will be attracted by all the insects that fly above the pond as well as the easy homes you provide for them.

Bat boxes on a pine tree in Martin Crawford's forest garden

Remember, ***Everything gardens***. All you have to do is identify who's out there who likes to feed on any particular species and invite them along to the party! This kind of biological control is an effective part of any Integrated Pest Management strategy.

Putting it all together...

So play around with placing systems and elements in your design until it feels like you've created the most effective layout. Start by placing everything in its ideal niche, thinking only about the needs of each element in relation to the microclimates on the site. Then consider how the elements might be better placed to integrate into systems, reducing work and pollution. The obvious solutions aren't always the best ones, hence the value of tools like *random assembly*. Desired outcomes will also determine the ideal method of integration.

One particular example involves ducks in a market garden. If our main purpose for including them is egg production, we can most easily keep them safe from predators by placing their duck house on an island in the middle of a pond (we'll need to have a strategy for collecting the eggs though). However, if we also wish to employ the ducks as part of a slug control system then we need to create a desire line for them through the vegetable garden. This we can do by locating the duck house and the pond on opposite sides of the garden, so every morning and evening our feathered friends pass through it, eating slugs as they go. Using this layout we create a flow of ducks, or slug-eaters, where we most need them. If they ultimately develop a taste for greens, we can feed them a 'permitted ration' each time before they pass through.

The pond could of course also be used to provide regular or emergency irrigation for the vegetables. If so, placing the garden down-slope of it will allow this to happen under gravity. Considering the way the pond itself is fed, may determine its own location. A pond can also reflect valuable extra light into a building during winter. Such a building may be your zone 0 focus around which everything else is located. This may seem like a complex process, but each relationship is a simple one. We just need to put them all together, prioritising the most important connections whenever a choice must be made.

Summary

Key things to remember about placement and integration are:

Placement

- *May start with a 'blank slate', or with one or more existing points of activity to arrange systems/elements around.*
- *Playing with a model, or moving 'element cards' around a plan is a good way to experiment with different locations.*
- *Always design from patterns to details.*
- *Use the key planning tools of zones, sectors and elevation to identify best placements of elements.*
- *Ensure that systems needing regular off-site inputs and those that create outputs to be exported off-site, are located near access routes.*
- *Make use of existing desire lines and create new ones where it's advantageous to do so.*
- *Remember soil type when designing planting schemes.*
- *McHarg's exclusion method can help with placing important or fussy elements.*
- *Visualise succession – remember that conditions on the site will change over time. Place elements appropriately.*

Integration

- *Once the best placements have been determined for each element, experiment with creating beneficial relationships.*
- *Look to create energy/resource cycles by using an output of one system as an input for another.*
- *Use natural patterns to direct energies across a site including principles such as stacking and succession.*
- *Seek out opportunities to take advantage of the edge effect, though not at the expense of clear desire lines.*
- *Use co-action analysis to maximise beneficial relationships and experiment with creating guilds.*
- *Use biological control to address imbalances.*

Integration flowchart

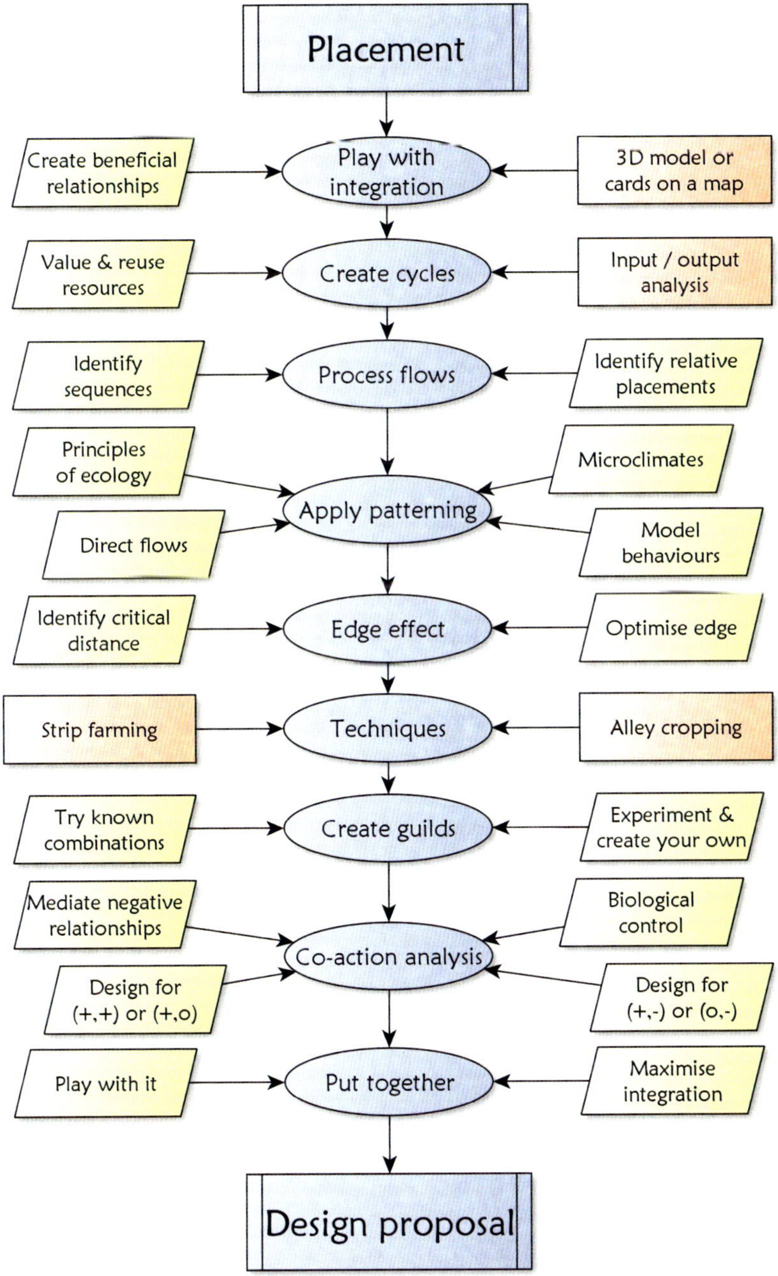

Design Proposal

After all our analysis, we must eventually commit to sharing our best ideas. No design is ever perfect, so don't be afraid to make decisions; you'll always learn from them, even if only with the benefit of hindsight.

Share your concept design early

Rather than waiting until you've filled in all the details, it's wise to share your main ideas with the clients once you have them. The feedback you receive will enable you to make any necessary adjustments and move confidently forward with the rest of your designing. This could be done by phone or email, but it would be better to meet up in person.

Use the opportunity to present your main ideas for discussion, perhaps using a model or map and cards to allow the clients to interact. The clients may love everything, but it's also possible that a vital piece of information was missed out during the interview that makes one or more of your ideas unsuitable. If any client expresses opposition to something, get to the root of the issue and see if any fine-tuning can satisfy them.

Remember that a client's perceived ownership of the design is an important factor in its success, so do your best to ensure they all feel heard and that their needs are being met. Your aim should be to obtain agreement about the main elements by the end of the meeting and to clarify any planning issues that would be involved. Ask at this point too if they've any preferences for people to work on the implementation.

Taking account of what you've learned, go back and fill in the finer details of the design to a level that seems appropriate. Ideally gather further feedback as you progress, but first clarify the level of involvement the clients want in this process. There may be times when all you have to work with is your initial interview, so if that's the case, make sure it's a good one!

Creating your final proposal

When you've reached the end of your analysis, it's time to put all your ideas together into a proposal for the client(s). We'll focus here on what might go into a written or printed report, but you'll also need to refer to these things if you make a final presentation to your clients. I offer a format for doing that later.

There are usually two main parts to a design proposal:

* *Maps (one or more) – of some form, showing spatial and possibly temporal relationships between elements.*
* *Supporting paperwork detailing additional information such as PASTE surveys, resource lists, designed guilds, planting plans, implementation and maintenance plans etc.*

Mapping your ideas

Exactly how you choose to map your ideas is up to you. Any format is fine as long as it conveys your design proposals in a way that they can be easily understood, by your clients and anyone else involved in the implementation process. Most commonly, ideas are presented as drawings, though you might choose to use a different method, such as ***CAD*** or ***modelling*** if you have those skills and/or they offer other advantages. The maps you create may need to be seen from a distance if you're making a presentation, so make them big enough if that's the case or ensure that you can project them instead.

A map can be defined as anything that indicates how systems and elements relate to each other,[28] so models, ***mind maps*** and process ***flowcharts*** can all be considered maps too. Just decide which is the most appropriate to use in each case.

What follows are a few examples provided for inspiration – some from my own Diploma design portfolio, others created by fellow designers. Each of these can be viewed in more detail in the online resources. A few more examples are offered in the section on designing other systems.

Drawing

Draw your design directly onto the base map

Perhaps our simplest option is to draw our design ideas directly onto our base map. If we made multiple copies of this earlier for recording site information, we can use one of those. If we wish to separate out some aspects of our design, or distinct implementation phases, we can avoid clutter by spreading these details over a series of maps. Below is the plan I made of my mobile home garden, drawn directly onto the base map.

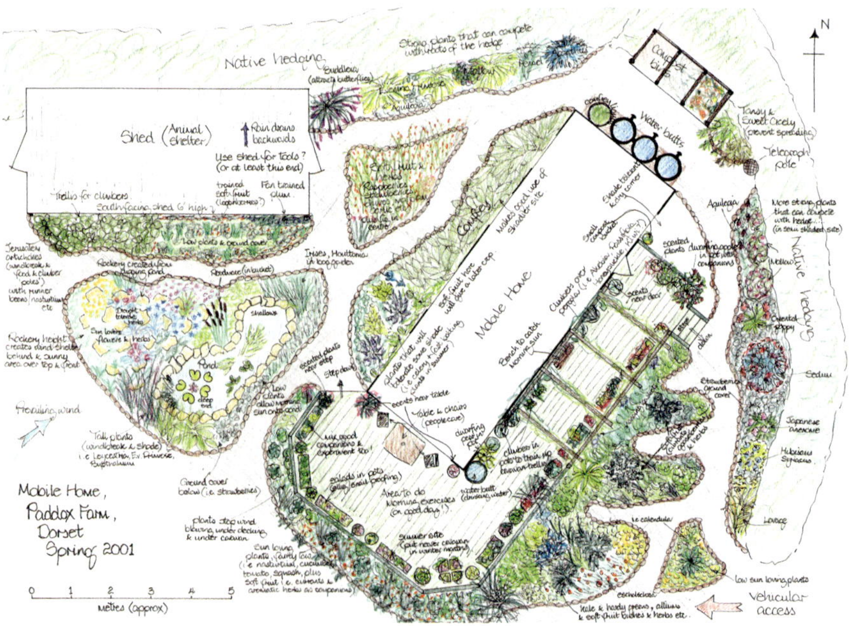

Design: Aranya

Base map and overlays

Another way of presenting a design is to instead draw it onto an overlay. If you chose earlier to record site information in this way, you can use the same technique to present your final ideas. Use good quality (high transparency) tracing paper if you can; this will also allow you to show multiple overlays at the same time when useful.

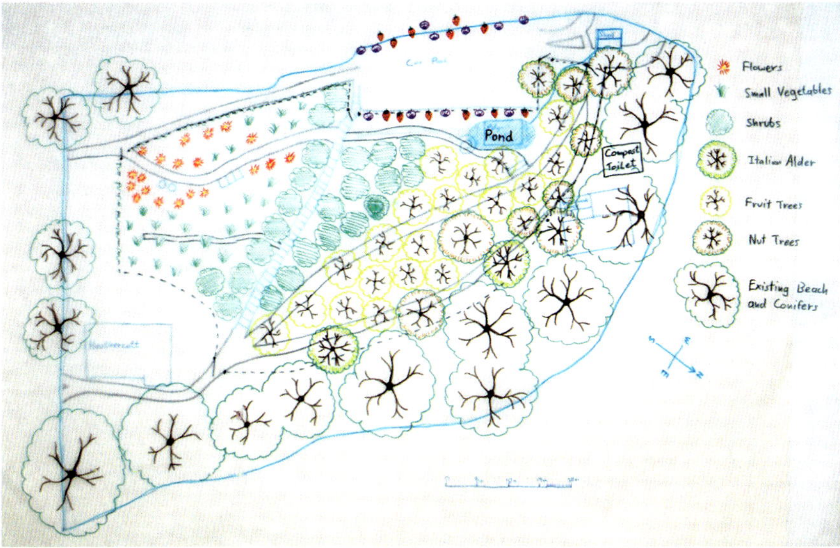

Design: Andrew Brackenborough / Colin Dunn / Sophie Ege / Randy Mayers

Overview plus highlighted details

If you're designing a large site with a need to convey more detail in particular places, you could scale up those areas inside little inset maps like this:

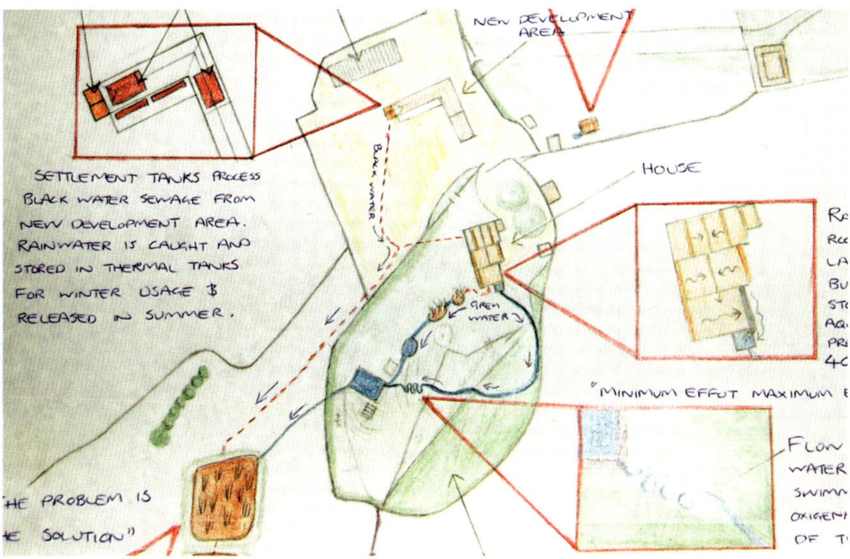

Design: Ellie Cooley / Gari Porteous

Design ideas on photos

Another way to convey ideas is to take photos of the site as it is and use either tracing paper overlays or paper cut outs to show any proposed new elements. Of course, if you have the skills this can also be done on a computer these days too.

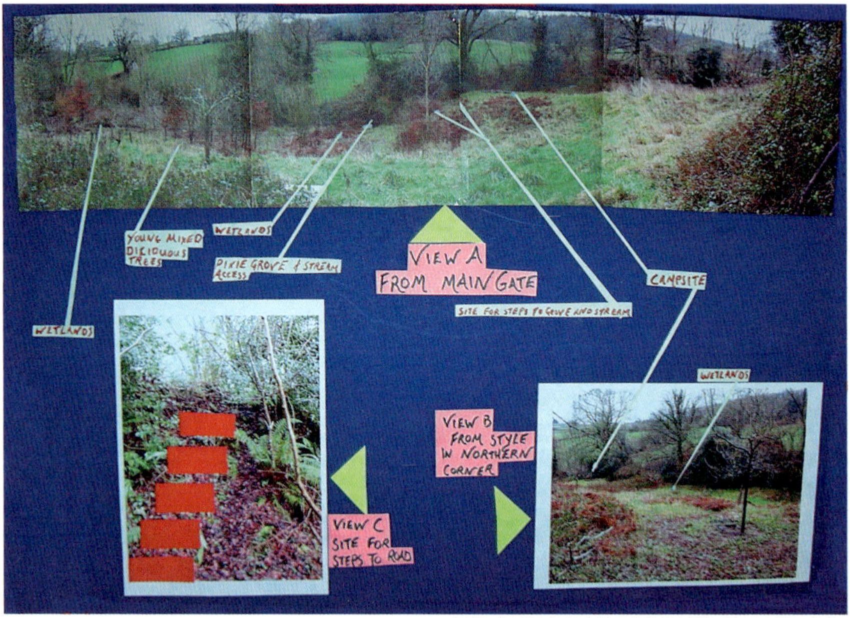

Design: Brendan Buesnel / Francis Burton / Kay Johnson / Tom Purling

Artistic cross sections

These are useful additions for showing relative heights and can also include information like approximate seasonal shade lines.

Design: Vinnie Blackwall

Artist's impressions

If you have drawing skills, these are helpful in giving your clients a sense of what your proposals will look like in 3D. Sometimes it can be worth the cost of employing an artist to do this for you, as a few images like this can make a big difference in presenting your ideas by bringing them to life.

Design: Ed Barsley / Stephen Noshie / Adam Rogerson / Marco daSilva

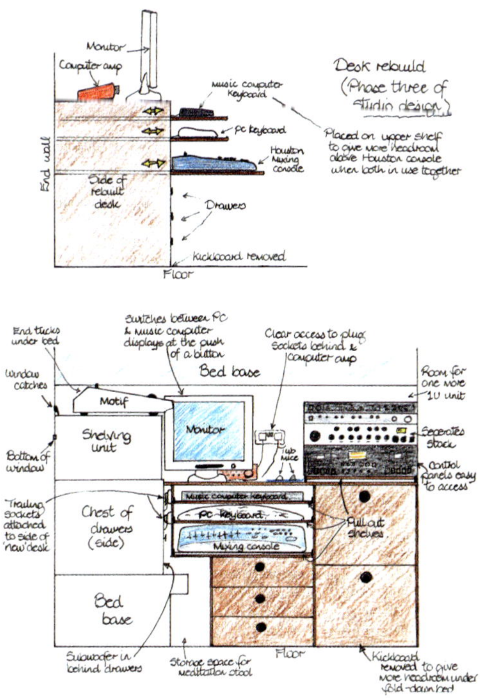

Design: Aranya

Front, plan and side elevations

Technical drawings can be helpful in providing any client with a clear perspective from three directions. In this example, a redesign for a small office space in my mobile home, we see phase three of an incremental design. As more functionality was required from the space, additional pieces of equipment had to be accommodated. This drawing shows the side and front elevations of the main desk layout.

Modelling

Adding a third dimension to your ideas

Modelling your design in 3D, even in a temporary media like sand, can help bring your ideas to life. This can be particularly useful when explaining water harvesting strategies and so on.[29] Or just add a 3D element to drawings like these examples:

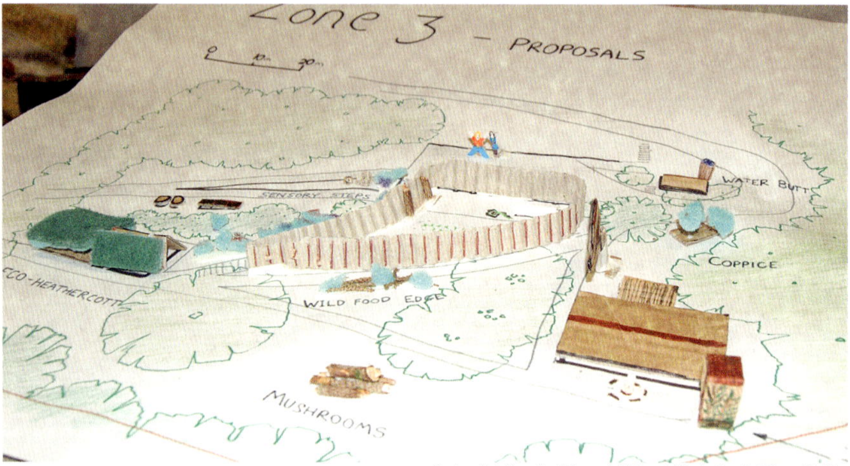

Design: Rob Knowles / Clement Malcurat / Lizzie Meade / Simon Watkins

Design: Elise Bijl / Danny Brown / Kate McGuire / Helen Smith

Using technology

Slideshow presentation

Technology can help if you have to deliver a presentation to a large group. Overhead projectors and 35mm transparencies are simpler but old technologies; a laptop and digital projector may involve more investment, but can often be borrowed or hired. While learning to use it takes time, software doesn't have to cost money. You may already own Powerpoint (as part of Microsoft Office), alternatively open source programs such as Impress, (in Open Office or Libre Office) can be downloaded for free via the Internet.

Computer aided design

If you have the skills, then CAD can be a good way to convey your ideas, especially given the built-in overlays. Regenerative Agriculture designer and practitioner Darren Doherty uses the computer to great effect in his work, here's one example:

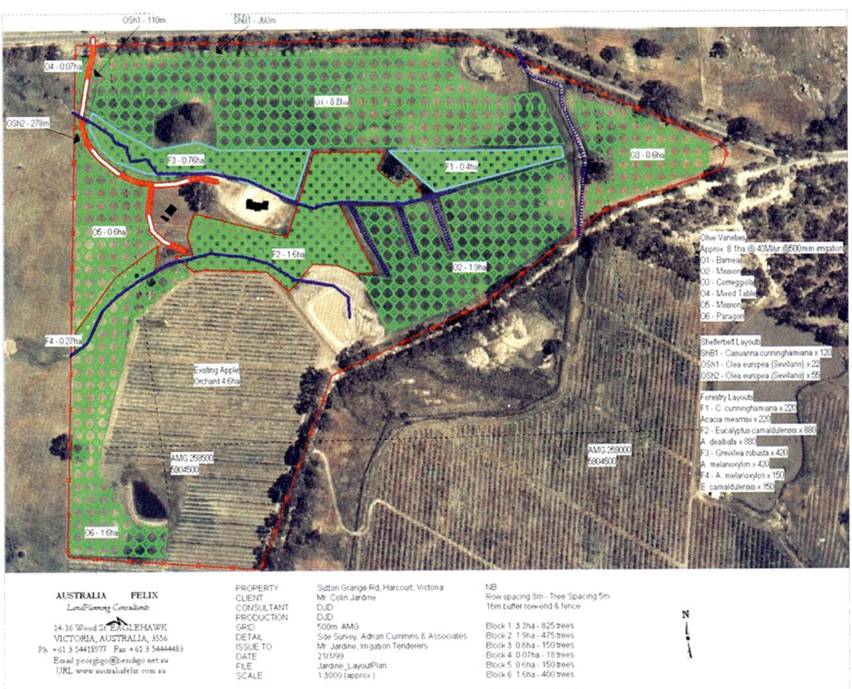

Heenan Doherty

Additional maps

In addition to your main site design proposal you might find it helpful to include other maps that show:

An overview of the area – perhaps showing specific aspects of the surrounding area, such as local settlements, access routes, soil, geology, watersheds etc.

Current site systems and elements – as maps or overlays on a base map, showing water, energy, vegetation, utilities, soil etc.

Zones, sectors, microclimates, desire lines, access points, boundary condition etc. – as maps or overlays to help clarify your design process.

Any transect maps that you surveyed and recorded.

Proposed re-zoning and any changes to access points – to improve current process flows.

Proposed systems – e.g. water, energy, production processes, showing fine details like water harvesting from buildings, or broad patterns like Keyline® earthworks.

Details of specific areas – maybe showing planting patterns.

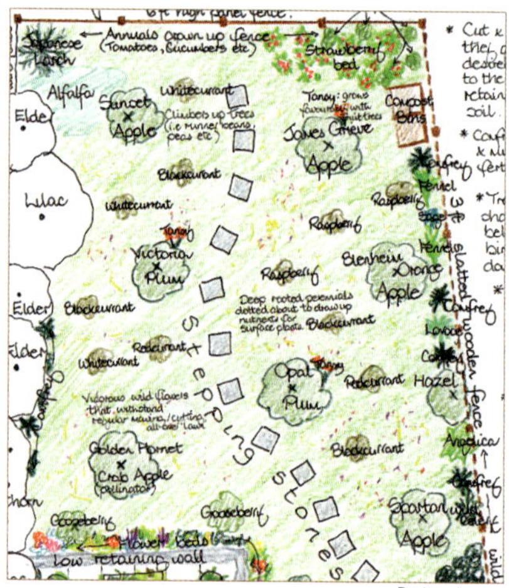

Design: Aranya

Further information

However you decide to map the spatial information, there will always be a limit to the amount of detail that you can include in any of the above. Here's a list of supporting documents that you might also want to include in your report.

A summary of client interviews – reflecting back their needs and wants, values and vision, resources and limiting factors, and their timescale for the design.

Themed details – specific reports on soil, water and energy systems as they currently stand.

A list of on-site and local resources – PASTE, DAFOR etc.

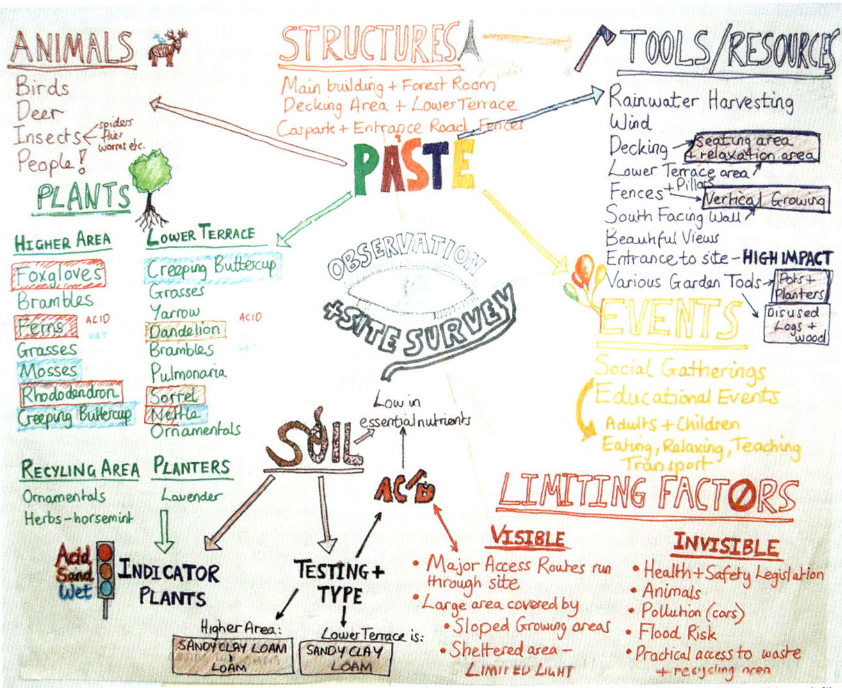

Design: Ed Barsley / Stephen Noshie / Adam Rogerson / Marco daSilva

Photos of:

- ***Energy or resource leaks*** such as erosion, visible loss of energy, water etc.
- ***Shade through the seasons.***
- ***Any important sight lines.***
- ***Any particular features you want to highlight.***

An overview of the main limiting factors – including any energy and resource leaks.

A list of key functions and any associated SMART goals.

An overview of your analysis:

- ***Any conflicting issues.***
- ***Your decision making processes*** – perhaps mentioning ethics, principles, any design tools etc.
- ***Any connections you made*** – creating patterns, flows or cycles.

Your recommendations:

* ***Overcoming limiting factors*** – *strategies to deal with the* primary restraints/leaks.
* ***Key systems*** – compare current patterns of use with your key proposed changes (zones, *desire lines*, etc.)
* ***Systems details*** – water, energy etc.
* ***Any decisions to be made?***

A list of required resources – those already available in place, those elsewhere on-site and those to import.

Information sheets – perhaps along with photos, to explain any unusual systems or techniques

A breakdown of finances – and estimate showing quantities and costs of elements and how these fall within budget.

A list of potential funding bodies – for any systems needing further financial input.

A list of potential partner organisations – e.g. the British Permaculture Association, via its LAND or Projects Networks.

A list of recommended species/varieties/breeds – any relevant flora, fungi or fauna. On a large site you'd be unlikely to create a planting layout for each bed, but you might provide a list of species, suggested guilds and perhaps some key plant positioning.

An Integrated Pest Management strategy.

Recommended books, courses or websites – for further reading and research.

Implementation and maintenance plans – we'll look at these in more detail next...

Recommended people – for the implementation team.

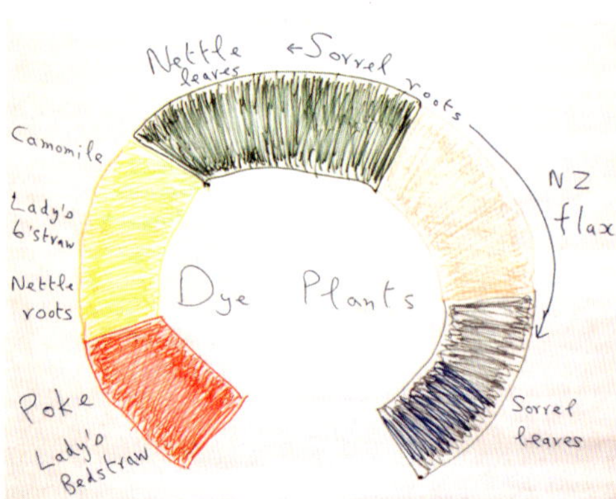

Design: Robin Aaronson / Liz Coates / Be Oliver

Implementation details

Your design report still needs a plan for implementing all your ideas, so having looked primarily at how we might lay out our design in space, we'll next consider the dimension of time.

An implementation plan may be simple or complex, depending on the scale of the design and who is involved. Mind maps are one way I've often used to show the main elements of a simple implementation; here's one example:

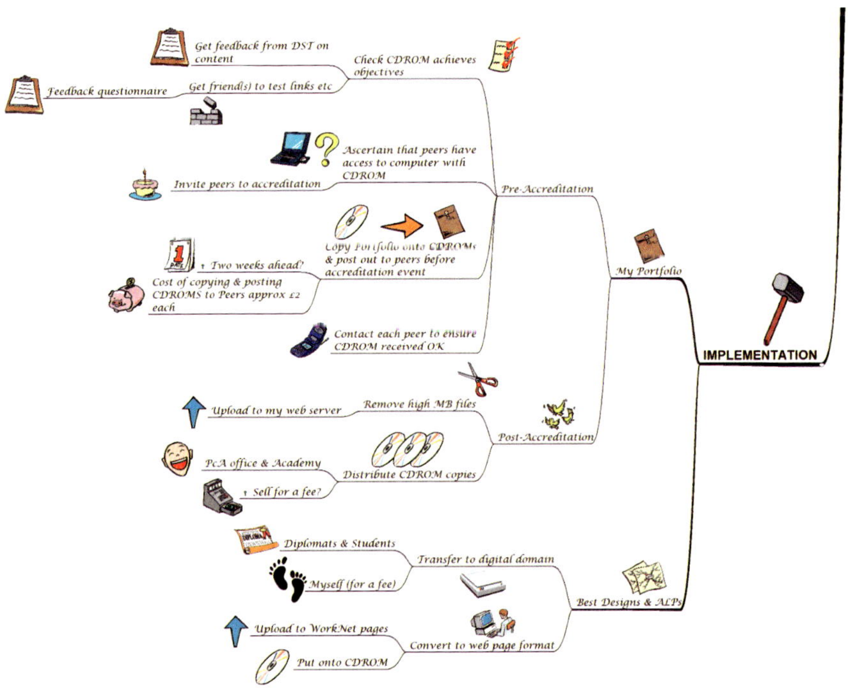

Design: Aranya

That said, most clients are going to need you to supply much more detail than this. In what order should they implement the design? Where should they start and how long might the work take? We'll consider how to do this next.

Summary

This section has been about committing your ideas to an appropriate format for communicating them to your clients. You don't need to have great artistic talent, though if you have it can help make your proposals look particularly attractive. The main thing is that you document your ideas in a way that can be easily understood and followed up on by your clients.

Accept that:

* *We all make mistakes and some of your ideas won't work, so don't let that stop you committing to action.*
* *A design proposal usually contains some maps, along with additional supporting documentation.*
* *If you're making a presentation as well as providing a report, you'll probably need maps and some other documents in a larger format for easier viewing.*
* *Use a mapping format that you feel comfortable with; as long as it communicates your ideas it's usually fine. Many of the examples shown here are particularly beautiful, but they really only need to be clear.*
* *If it's a particularly important job, it might be worth employing an artist to convey your ideas for you.*
* *Models, mind maps and flowcharts are maps too.*
* *Technology can sometimes make life easier, but can also leave you reliant on its good functioning.*
* *Define the functions and SMART goals you've been designing to achieve.*
* *Don't be afraid to include details of your analysis; it will help your clients understand the design process.*
* *Include extra maps of specific systems for clarity.*
* *Use photos to help illustrate any important issues and for showing proposed changes.*
* *Only include details of systems that you're familiar with, otherwise refer clients to an expert in that specific field.*
* *Refer your clients to where they can learn more about permaculture or any specific elements of the design.*

Design proposal flowchart

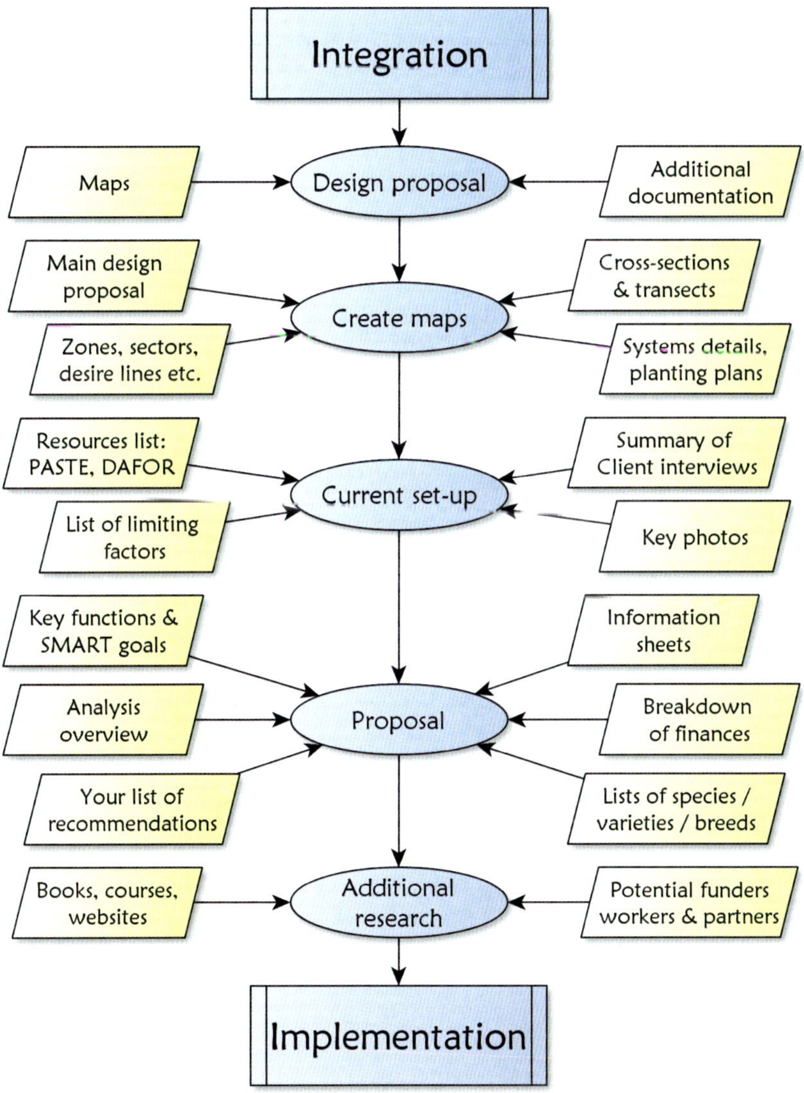

Implementation

So we are almost at the point now where we can impose our design ideas upon the real world, but first we need to create an implementation plan to guide us.

Breaking down the task

We'll start by breaking down the whole process into its major sub-tasks, creating more manageable chunks. These may already be very clear from our design process. For a simple design, or where you are both designer and client, this may be adequate enough for you to work from. For more complex designs, you may already have created a design around specific phases, each one with its own budget, deadline or SMART goal, and map. Whatever degree of complexity we start with, we'll want to divide up each sub-task to create a ***Work Breakdown Structure*** (WBS).

Start small and work out from well-managed areas

Set realistic goals – many small successes make you and the client feel good and create added enthusiasm for the project. As you can see, the WBS forms a branching pattern, starting with the end objective at the top.

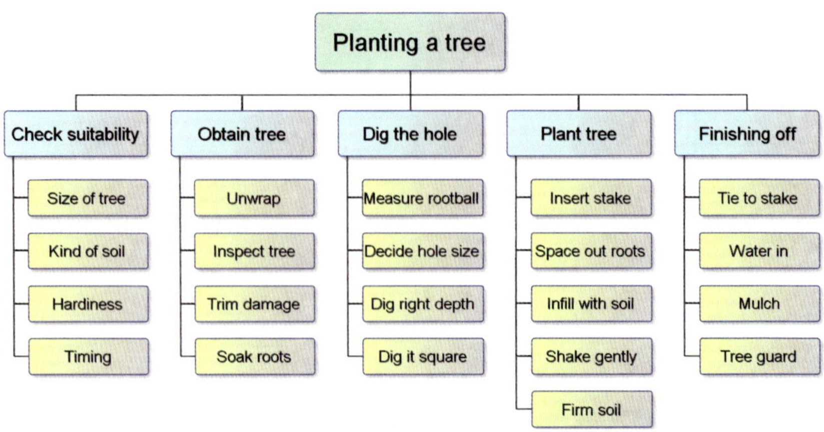

A simple Work Breakdown Structure for planting a tree

Sub-tasks are then connected to this and in turn divided up into smaller tasks, components and so on, until each part is a manageable size. This forms the basis of our plan, though a WBS doesn't identify priorities or the timing of activities.

Factors that influence task prioritisation

In order to start prioritising tasks, we need to consider all the factors that might influence this. The following can help you decide what should be done when, and identify where certain tasks depend upon the completion of others. I usually consider them in this order:

Least change for greatest effect

What little thing can you do that will create the most benefit from the time or resources invested? Of course, what constitutes 'little' will vary depending upon the scale of the project. Often this might be plugging a leak, such as a regular loss of soil, money, or energy. Building soils for instance can reduce the need for irrigation, which can save money on earthworks or other irrigation infrastructure. It's always easier to conserve energy than use up resources to generate more. Small investments that save you money or resources in the longer term (e.g. building insulation) can free up additional funds to make bigger money-saving investments in the future. Early successes can encourage further action, so give your client an instant payback and let them see the value of your design advice.

Design around the limiting factors

Seasonal constraints must be considered when planting trees or sowing crops. If trees are required and money is also a limiting factor, bare-rooted trees become a good investment. They are cheaper to buy than pot grown trees, but can only be planted safely during the colder winter months, so two limiting factors combine to give high priority to winter tree planting. Likewise, if you want a good harvest from your garden in the autumn, there's a fairly small window of time (in the spring) in which to sow most of your crops. Seeds are cheaper than plants. Remember McHarg's exclusion method (p122)? We can also apply the concept to time. Identify your most important or most difficult or time-constrained tasks and then put them onto your calendar first.

Resource availability

Many limiting factors can become surpluses at other times. For many people with a land-based income, money is much more abundant in the summer and autumn than in the winter. Conversely though, time may become a key resource when there's a lot less to do in the garden. Jobs that require a lot of finance or labour should be done when these surpluses are available. A surplus of labour may also bring extra skills. If the site is suitable, running courses is a good way to take advantage of this. As well as potentially providing a financial yield, a practical course can also bring in an expert and a lot of keen volunteer labour to get a big project done.

Dependencies (a.k.a. precedences)

There are certain things that need to be completed, or at least part-completed, before others can be started. A building needs its foundations laid before its roof is put on, or to make a sandwich you first need to bake the bread. These are called causal dependencies. However there are other factors that can determine the pace at which a project can proceed. Sometimes the availability of resources is the key limiting factor and sometimes it simply makes sense from an organisational perspective to arrange tasks in a certain order. For example, you may have a particularly skilled person on site, who can do two distinct tasks not normally implemented at the same time. These are discretionary dependencies. The key thing of course is to remember to place any dependent activities on the timeline after those that they are dependent upon.

Yeoman's Keyline scale of permanence[30]

Remember that when working on a broadscale site, Yeoman's scale lays out the relative permanence of the main elements you'll be working with.

1. Climate 2. Landform 3. Water supply 4. Roads
5. Plant systems 6. Microclimates 7. Buildings
8. Subdivisional fences 9. Soil

This suggests putting your water systems in first, then access roads, then trees and so on. However, subsoil ploughing along with good grazing management can rapidly increase the depth of soils and change the irrigation needs of a landscape. Thus, soil restoration needs to preceed significant dam building.

Creating a simple implementation plan

If you're both the designer and the client, such a plan will only have to make sense to you, in which case it could be a relatively simple affair, such as a basic list of tasks set out on a timeline. In this situation, without any additional labour or skills being involved, you potentially have as much flexibility as your own patience levels can stand and you only need to create a basic pattern to work from. This will still require an understanding of dependencies, but not the finer details around timings.

I find a simple, but useful tool for designing such processes is to use ***pieces of paper***, or ***sticky notes*** that I can move around on a timeline or calendar until I'm happy with the ordering. All you're aiming to do is to create a basic pattern, identifying the order in which tasks will need to be performed. The finer details of particular tasks can be identified elsewhere.

A simple timetable can certainly be good enough if you're not working with others, but as soon as you start designing with or for someone else, you'll need to come up with something a bit more substantial, that is both thorough and clear. And remember too that skilled people are rarely available at short notice. You'll need to identify where in the process any will be needed, so you can confidently book them far enough in advance.

Design: William Baker Morrison / Ailbhe Gerrard / Xavier de la Huerga / Rebecca Rothera

Creating a project management plan

Luckily for us, people have been managing projects for a long time and there are already a number of useful tools to help us put such a plan together.

Either you'll be creating a plan for someone else to manage, or you'll be overseeing it yourself. Both situations will have a start and an end point for you, which you should aim to define, either in terms of time or task completion. You and the client(s) should be clear about your own degrees of involvement and when you hand over full responsibility to them and walk away.

A successful project needs to meet the client's requirements, be within budget, and be completed on time. You should be clear before you start about what these criteria are and create a plan to give you the best possible chance of meeting them.

As designer, you're responsible for reducing the project to a series of manageable tasks (WBS) with SMART goals, identifying sources for any required resources, and for building the team to do the work. This would include finding a project manager to oversee the implementation phase should you hand over responsibility after producing the plan.

If you remain to oversee the project through to completion, you'll also need to have good financial management and people skills. You'll probably also need to be a good problem solver, as projects rarely go completely to plan.

To create a realistic schedule you'll need a certain amount of knowledge about the time, energy and resources required to achieve each task. If you're not sure about any of these things, to ensure your figures are reasonable, ask those who are likely to be doing the job. Resources are often easier delivered to a site as required, to reduce the need for storage, but remember that delays can occur and prices can go up...

The kind of things that can go wrong include:

- *Poor communication, the team don't know what to do*
- *Disagreements, personality conflicts and walk outs*
- *Failure to meet standards or regulations*
- *Bad management and poorly defined goals*
- *Inclement weather*

Being prepared for such eventualities gives you a better chance of dealing with them effectively should they occur.

A good tool for taking a simple Work Breakdown Structure and turning it into something more precise that we can work from is a ***Gantt project planning chart***.

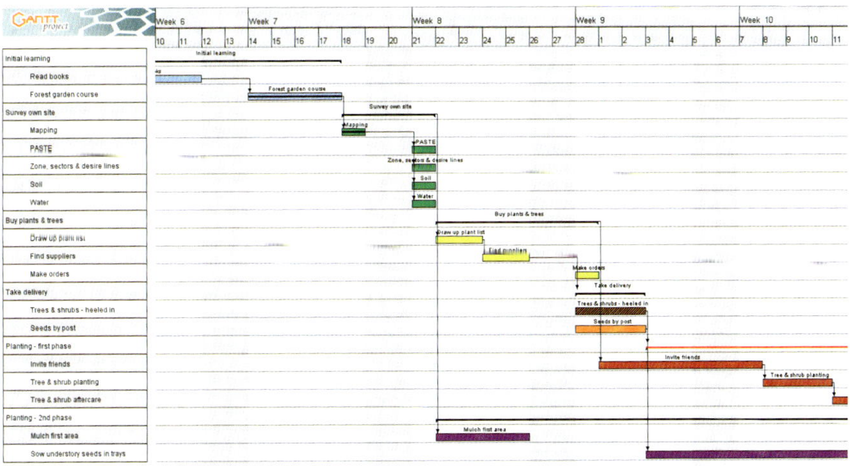

Gantt chart created using Open Source software, though you could also use paper to lay one out.

One advantage of using software is that if a task is delayed, then all dependencies can be easily shifted along the timeline in response. If you are computer-literate, the best way to learn how to use this useful tool is to download one of the free Gantt programs and play with what it can do. A Gantt chart places all the chunked tasks down the left side of a grid, in the order of implementation. The timeline runs left to right, and for each task a coloured bar is placed indicating the anticipated start and finish times.

Where one task depends upon the completion of another, an arrow links them. This provides a clear visual picture of the relationship between tasks and allows you to see which ones are critical to the whole project being finished on time.

Critical Path Analysis

If you have a tight deadline and are working with multiple contractors, you might need to do a ***Critical Path Analysis***. This process calculates the longest path (in terms of time) of all planned activities to the end of the project, and tells you the earliest and latest that each activity can start and finish without making the project any longer. If maths is not your thing, you might want to skip the next paragraph!

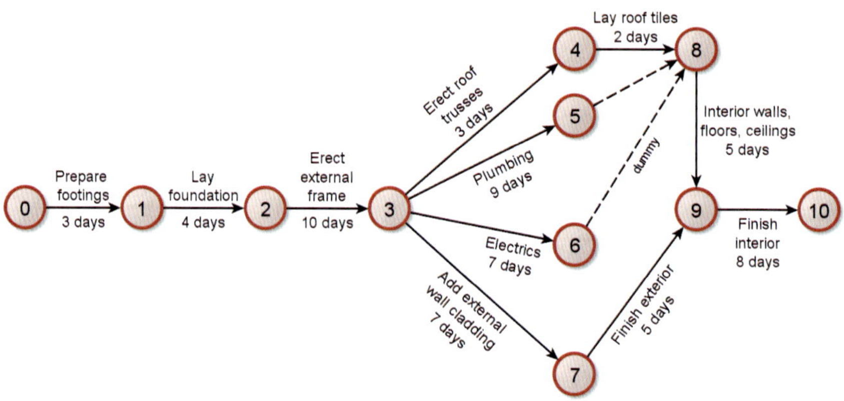

Our example above is called a CPM/PERT† diagram. Here the longest path is 0-1-2-3 (each dependent upon the other), then 5-8-9-10. This is because 3-5-8-9 takes 14 days, two more than 3-7-9 or 3-6-8-9, both at 12 days and four more than 3-4-8-9 which takes only 10 days. Therefore our critical path is 0-1-2-3-5-8-9-10 and is 39 days long. Any delays along this pathway will delay the whole project, whereas tasks not on the critical path have some leeway, for instance 3-4-8 can overrun by four days and still not delay proceedings overall. Knowing this shows us which projects may need extra people bringing in if they get behind and which can be given the flexibility to deliver a little late and not affect the overall project deadline.

† *Critical Path Method / Project Evaluation and Review Technique.*

Again, you can draw one of these diagrams out by hand, but Gantt chart software can usually create one of these for you automatically, saving you time. If, in addition to the most likely time for a task to complete, you enter into the software your estimates of the most optimistic and pessimistic timings, it can provide you with earliest and latest start and finish times (EST, EFT, LST and LFT) for the project too.

Overseeing a project implementation

There's only room here for an overview of this process, so if you do find yourself about to oversee a project of any reasonable size, I'd recommend seeking out at least a good book, if not training first. While mistakes made on small project implementations are unlikely to be serious, when working on anything of significant budget and involving contractors, you really want to be well prepared for any eventuality.

That said, with your SMART goals, Gantt chart and CPM/PERT diagrams, you're half way to successfully overseeing the project. One key thing to remember is that you're working with a system and as such it may well behave unexpectedly. That system might include multiple providers/contractors, so a good project manager is one that can manage and motivate their team to complete on time. Indeed any of those things mentioned previously could go wrong, so foresight and adaptability on your part are important.

Not all of us feel enthused about working with numbers, but managing cash flow is another important skill in project management. It's one thing to plan out what will happen when, but we also need to ensure that tools, resources and contractors are paid on time (and that includes you!), to ensure they do. As an example of what can go wrong, a friend of mine once walked away from an exciting project in Spain, because his boss didn't make available the money required to enable him to do his job. Don't forget either that money can take time to travel between bank accounts.

Your ability to anticipate possible scenarios will help you to be prepared when things do go awry. If a particular task looks like it may take longer than planned, you can refer to your Gantt chart or CPM/PERT diagram to find out if it's a critical task (i.e. will delay the whole project if it overruns). If it is, you can then assign extra people to it to ensure it, and the project as a whole finishes on schedule. If it's not critical, you'll be reassured by how much slack time you still have in hand.

Another issue that can occur, particularly on longer-term projects, is a change in budget. We might be less worried about having extra money to spend, than suddenly having the budget cut. This is where some flexibility and those 'Plan B's come in. You did include those in your design didn't you? Being well prepared for any realistic eventuality is the key to successful project management. I wish you well in creating a design that is beautiful, sustainable and functional. However, even after you walk away there will still be plenty to do and you'll need to train your client to monitor the performance of the design over time and to ideally make some improvements. This can be an in-built part of a maintenance plan and we'll look at how to create one of those next.

Implemented garden design around my former mobile home

Summary

The key things to remember about implementation are:

Breaking down and prioritising tasks

* Create a Work Breakdown Structure by chunking large tasks into smaller, more easily tackled sub-tasks.
* Identify those small changes that can have an immediate significant beneficial effect.
* Notice which tasks are limited by seasonal constraints and place them on the calendar next.
* Determine which tasks need a lot of labour or resources and plan to do them when they are abundantly available.
* Identify dependencies and order those tasks accordingly.

Creating a plan

* A simple plan may be enough when you're also the client.
* Sticky notes or sheets or paper are a useful planning tool.
* Spreadsheets can be useful too.

Project management

* Create a project management plan for significantly big implementations or when designing for someone else.
* Clarify any SMART goals, as they'll define your timescale and in turn how much inside your budget you will be.
* Determine your own and the client's levels of involvement in the implementation process.
* Create a realistic schedule. Use individual task timings (refer to the contractors if you are unsure) and what you know about task dependencies, to create a Gantt chart.
* Identify the critical path and prioritise those tasks along it.
* Build and motivate the team to implement the design.
* Things can, and probably will, go wrong. Be prepared.
* Get more training if you're doing this for anything other than a small project.

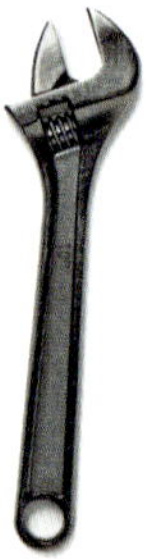

Maintenance and Evaluation

The ideal that we aim towards in permaculture is the concept of '*harvesting as maintenance*'. In reality, we've done pretty well if we create a design that requires less inputs over time while also producing increasing yields.

Creating a maintenance plan

As the implementation of a design progresses, the time and resources used in ongoing maintenance have to be subtracted from those available for further development. So where you have a choice implement the lowest maintenance systems first and the highest last. This would allow your design to be completed in the shortest time. However, as we've seen in the previous section, sometimes our implementation is limited by the availability of labour or resources and can't always follow this ideal.

We're aiming to design a system needing as little maintenance as possible, but for intensive systems there may be a need to employ someone. Here are some other things to consider:

Life never stops

Plants and trees don't stay the same size and need some form of ongoing attention, even if only an annual prune and harvesting. At least the times for these tasks are often seasonal and so can be listed on a perpetual calendar.

Identify the ongoing availability of labour and resources

This will help you plan bigger maintenance tasks to coincide with these peaks. Being able to estimate the amount of time these jobs will take is also a useful skill to develop.

A stitch in time saves nine

The wear and tear of most non-living things is inevitable, but timely maintenance can save a heap of extra work later on. Include room in your plan for a regular check up of all systems and encourage the fixing of any small issues promptly.

Maintaining correct functioning

You may need to include mini maintenance plans to ensure the correct functioning of certain systems. Any unusual technologies, such as a tree bog composting toilet, might need clear instructions posted up to ensure that it continues to function as it should. Remember that if such systems fail to perform as intended (e.g. a compost loo smelling), then they may be abandoned as ineffective, perhaps forever colouring the user's opinion of the technology. Don't forget the concept of zoning either – stick the toilet instructions on the back of the door where the sitting user can read them.

The importance of a clear maintenance plan, available to every user

A system is only as good as an operator's understanding. In your absence, a clear and detailed maintenance plan is the next best thing. Make sure, before you leave, that clients understand what it all means and that everyone else interacting with the design has access to any information that is relevant to them.

The need for ongoing evaluation

Incremental design

Remember that not all your design decisions will be right for the site; some species might not survive, others the client may not like. Succession can cause microclimates across the site to move. The client's needs themselves may change. Any of these can stimulate a re-evaluation of parts of the design. The more familiar your clients are with the permaculture design process, the better able they'll be to make effective changes themselves.

Summary

Remember that no design is ever finished. The nicely functioning system you leave behind may gradually deteriorate without a good clear maintenance plan for the clients.

The key things to remember about maintenance planning are:

* *Aim to implement low-maintenance systems first, as they will take less time away from the establishment of the rest of the design.*
* *Time large maintenance tasks to coincide with peaks of labour or resource availability.*
* *Things last a lot longer given a little regular care and attention. Timetable in monthly checks of key systems.*
* *Provide instructions for unusual technologies to prevent their malfunction and abandonment.*
* *Ensure all site users have access to a clear site-wide maintenance plan.*

Evaluation and incremental design:

* *Designs rarely function completely as intended; to anticipate all the things that might happen would require an extensive design process. Ongoing evaluation is vital to ensure that any unexpected occurrences are quickly resolved. Almost all of the most successful designs have developed incrementally.*
* *Once you've left, the client will have to respond themselves to any unexpected issues. Do your best to ensure that they understand the design process and at least the basics of permaculture, so that they can care for the site effectively. Better still get them onto a design course.*

Implementation and maintenance flowchart

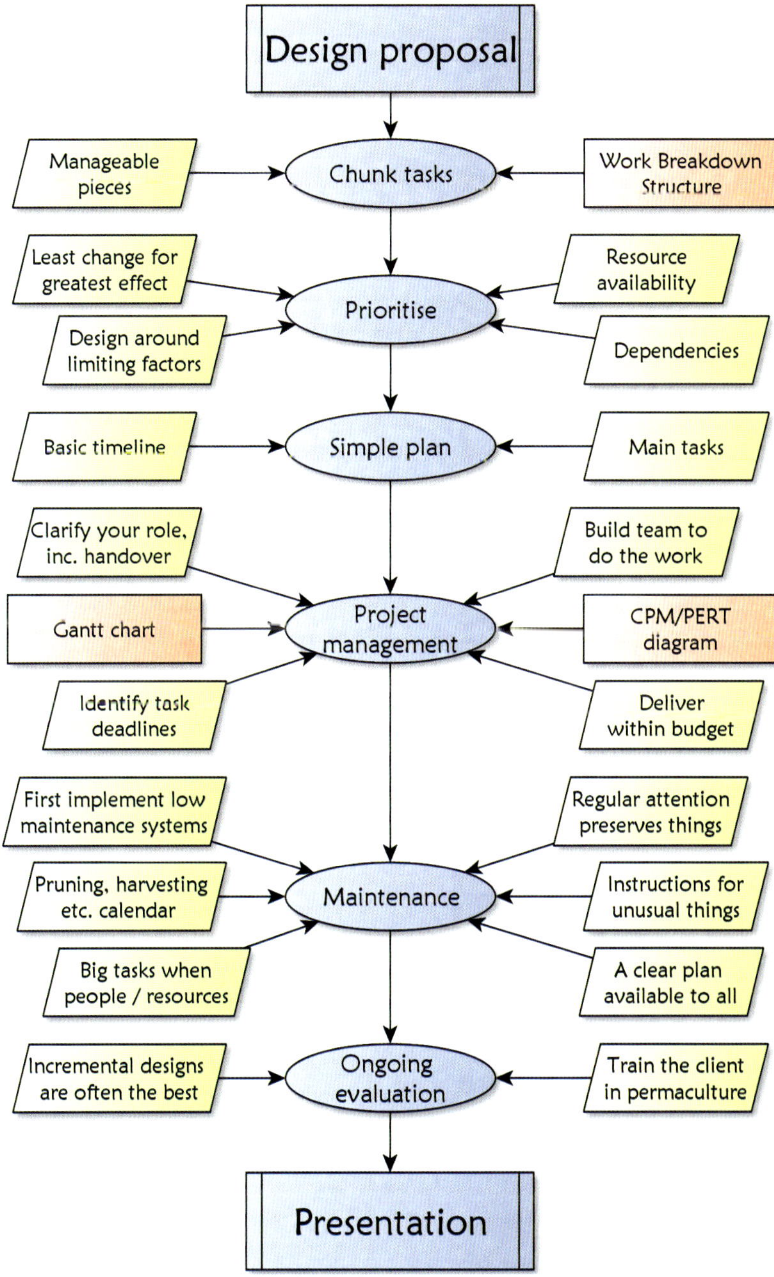

Presenting to a Client

Now that you've finished your design, you may be faced with presenting your ideas to the client(s). The first thing to remember is that however well they know the site, they won't have been through the same design process as you and won't be familiar with many of the analysis tools. Your job then is to communicate not just your recommendations, but also the reasoning behind the choices you've made. Sometimes you'll disagree with what they initially asked you for; if so, show good reasons, so that they can see why your ideas are a much better choice. Have at least one trial run presentation, particularly if you're not a confident public speaker. In my experience though, knowing the subject is the best cure for any pre-presentation nerves.

Observations

Describe your key site observations.

* Explain what you've discovered about the overall landscape surrounding the site (a map can help).
* Present your base map with any overlays showing zones, sectors, access points, desire lines etc.
* Describe your observations about slope across the site.
* What did you notice about microclimates across the site?
* What is the soil like? Does it vary in content, depth, pH etc. across the site?
* How does water flow and settle across the site?
* What significant flora, fungi and fauna are present?
* What are your observations about structures on the site?
* How is energy being harvested and used?
* How are any natural or social events affecting the site?
* Use photos where necessary to illustrate these points.

The Client interview(s)

* What are the key points you took from the interview(s)?
* Reflect back on their values and vision.
* Reflect back their timescale and budget for the design.

Boundaries / limiting factors

- What are the key limiting factors of the site?
- What energy and resource leaks have you noticed on site?
- What other limitations have you taken into account?
- What are the key limiting factors for the client(s)?
- What key restraints (physical/'invisible') of the client(s) have you taken account of in your design?

Resources

- What and who will be useful on, or local to the site?
  - What are the key resources that the site has abundantly?
  - What other resources (cheap, free, or plentiful) and skills have you identified as being locally available?
- What knowledge/experience/skills do the client(s) have?
  - What are the clients going to be able to contribute to the design (remember the importance of ownership)?

Analysis

- What did you identify as the key functions?
  - And what aspects of your client interview and site observation led you to choose these?
  - Frame these as SMART goals that you're aiming to meet.
  - What systems and elements did you choose to fulfil all those functions? And why did you choose them?
  - Explain a little about the analysis methods you used to help you make your decisions.
- What connections did you make?
  - What elements were you able to connect into mutually supported systems (or systems into patterns)?
- What principles/patterning did you apply?
  - Where did you apply permaculture principles/patterning to your design decisions? Explain the value of using natural patterns.

Design

Show your final design proposal.

- Reveal your final design drawings/overlays/model, or whatever format you decided upon to convey your ideas.
- Explain your main recommendations, the key things that will make the whole system function more effectively.

* Keep to your main points; mention where finer details, such as planting plans, are included in your report.
* Highlight any new or improved systems, such as water, energy or production cycles.
* Reassure your clients about unfamiliar technologies by letting them know that maintenance plans are included.
* Mention any potential funders or partner organisations.

Implementation

* Where should the client start?
    - Recommend at least one 'least effort for greatest effect' strategy that will ensure early success and motivate the client to do more.

Present your recommended implementation plan.

* Lay out a timeline with suggested tasks, ideally in a project management format, such as a Gantt chart.
* Explain the reasoning behind your prioritisation.
* Show how you have kept the design within budget.
* Offer some low, medium and high investment (time/cost) alternatives in case of any unexpected short, medium or long-term changes in income.
* Mention any contractors you'd recommend for the implementation.

Maintenance

Describe any maintenance required.

* What ongoing maintenance will need to be attended to? Mention any specific technologies needing special attention.
* Explain that a design is never finished and that part of the maintenance will involve ongoing monitoring and some occasional tweaking.

Questions

Invite any questions. Be honest. If you don't know the answer to something, say so, then offer to find out.

Present your report

Hand over the paperwork.

* Provide your clients with the whole proposal as a report, even if you are going to manage the implementation process yourself.

Presentation flowchart

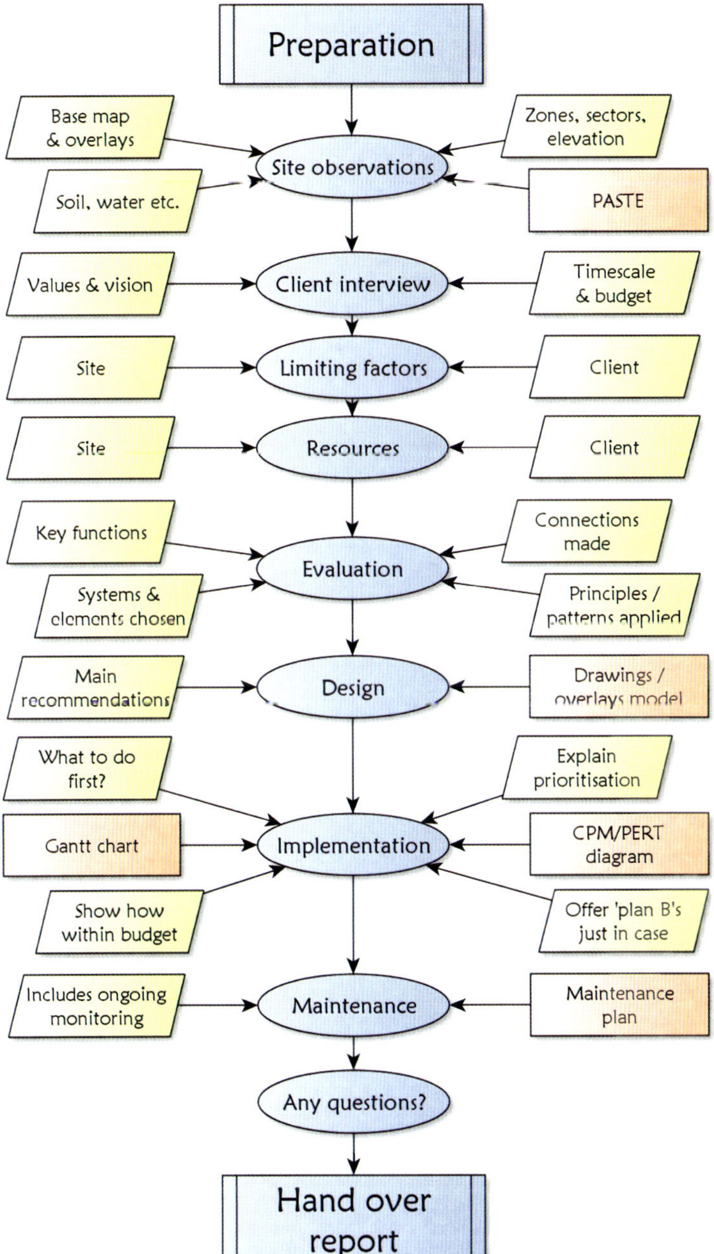

PART THREE

~

Beyond Land Based Design

Beyond Land Design

While permaculture was originally created to address the issues of unsustainable land use, over the last thirty years or so many have looked at how the same natural principles can be applied to improve the design of many of the other things that we do. There's only room here for a cursory look at all the possibilities (that's probably a whole other book in itself), but I'll give a few examples, some of which are from my Diploma in Applied Permaculture Design portfolio,[31] to help you start experimenting for yourself.

Designing processes

Up to now we've been considering how to design processes that harvest energy for us on an ongoing basis and then decide the best way in which to implement them. Indeed many of the things we do in life are processes: taking a journey, making a meal, or even writing a book, so perhaps it's not surprising that we can use those same principles to help us design them too. One process I'm particularly familiar with is organising courses. Using the same observation strategies as before, I firstly find out from potential clients what kind of courses they want me to run, and then I survey venues for suitability. Over the years, using ***incremental design***, I've developed my own well-oiled convening process. That said, there's always some adjustment needed to account for availability of fellow tutors or those all -important inspirational sites to visit.

When it comes to planning course timetables I find the sticky note technique useful again. In the Teacher Training course timetable shown opposite, we (the teachers) had each decided the content of our individual sessions, but still had to agree their best sequence. Being able to try out different placements by moving notes about really helped us. Of course, all these sticky notes were the same size, even though sessions were to be of different durations, so we wrote times on each one and checked our totals to make sure everything added up ok. A bit of maths is always involved, but this tool has been useful for us in planning course timetables together for many years.

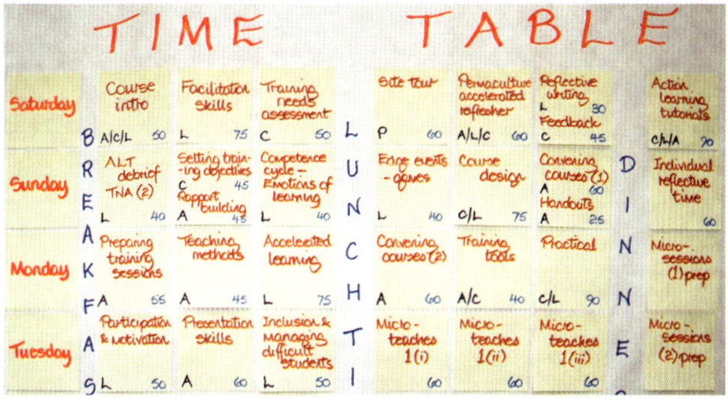

Sticky note timetable

Sticky notes on a large sheet of paper can even serve as your final version; it offers your timetable some dynamic flexibility when things don't go quite according to schedule, though the stickiness can wear off and it's vulnerable to a bit of mischievous tampering!

A more technological version of this tool is the ***spreadsheet***, which is good for more than just doing accounts. I use one of these when I can for planning my own course timetables, as it gives me flexibility to have blocks equating to the actual length of each session. It looks neater too and can be printed out so everyone gets a copy. The downside of this of course is the need for a computer to create it on and the extra investment of time spent learning how to make best use of the software.

Heathercombe Permaculture Design Course April 2011

Date	08:30	09:00	09:30–11:00	11:00	11:30–13:00	13:00–14:30	14:30–16:00	16:00	17:00–18:30	18:30–19:30	19:30–21:00
Session	Tidy up	Morning Circle	SESSION 1	BREAK	SESSION 2	LUNCH	SESSION 3	BREAK	SESSION 4	DINNER	SESSION 5
Sat 2/4							ARRIVALS Settling in, cup of tea etc		Site tour, Locating Ourselves	Domestics	Welcome, Opening Circle, Course Intro & Culture VIDEO: Greening the desert
Sun 3/4	Tidy up	Morning Circle	Introduction to Permaculture, Ethics & Principles		Intro to Design Process Needs & Yields analysis		Site observation		Pattern understanding / SLIDES: Patterns in nature		Accelerated learning
Mon 4/4	Tidy up	Morning Circle	Introduction to systems Problems & spirals of erosion		Principles of ecology walk		Wholistic site analysis Elevation planning		Wholistic site assessment exercise Biotime diaries, Intro to microclimates		SLIDES: Permaculture principles
Tues 5/4	Tidy up	Morning Circle	Review of fundamentals		Wild design game		FREE TIME / Optional eco-shopping trip to Proper Job recycling project				SLIDES: Natural building
Wed 6/4	Tidy up	Morning Circle	To Steward Wood / SITE VISIT: Steward Community Woodland site tour	Low impact planning	Practical		Whole community planning & design		Community issues, Group process		Action learning guilds
Thur 7/4	Tidy up	Morning Circle	Soil Intro: Fertility factors, types & textures		Soil: Fertility cycles, food web Tilling pros & cons		Site Surveying: Transect Soil profiles, texture / jar tests, Indicator plants		Soil conservation, remediation & improvement Biological interventions		SLIDES: Soil conservation & improvement
Fri 8/4	Tidy up	Morning Circle	Water introduction, characteristics & issues		Water: Find it, Measure it, Move it, Store it, Clean it, Use it		Rainwater harvesting - earthworks & tanks Water audit exercise		Aquaculture / Grey & black water, humanure, composting toilets		Skill sharing
Sat 9/4							FREE TIME				
Sun 10/4	Tidy up	Morning Circle	Midway review, Evaluation	Group process Design process intro & design areas walk	Design Work: Site survey		Site surveying tools: A-frame, bunyip, pacing	Design Work: Site survey (continued)	Maps & map making		Talking circle
Mon 11/4	Tidy up	Morning Circle	Design Work: Creating base map		Design Work: Zones & Sectors overlays		Design Work: Client interview	Design Work: Identify functions Choose systems & elements	Trees & their microclimates, Agroforestry		SLIDES: Agroforestry & forest gardening
Tues 12/4	To Landmatters		SITE VISIT: Landmatters site visit			Lunch at Landmatters / To ART	SITE VISIT: Agroforestry Research Trust Forest Garden site tour	Travel back to Heathercombe	Fruit trees & coppice		Zone 00
Wed 13/4	Tidy up	Morning Circle	Buildings & structures		Urban strategies (physical & invisible) Transition culture		FREE TIME				
Thur 14/4	Tidy up	Morning Circle	Design Work: Deciding placements		Integrated pest management		Foraging / Seed saving		Kitchen gardening / Cultivated ecology		SLIDES: Example designs Design Work: Deciding placements cont'd
Fri 15/4	Tidy up	Morning Circle	Real wealth & wiser money		Food choices		Design Work: Client feedback Design overlays		Design Work: Implementation & maintenance planning Finalising designs		Design Work: Preparing plans & notes
Sat 16/4		Group presentations & feedback	Group presentations & feedback		Group presentations & feedback		Feedback on process. Next steps: PAB, Networks, Diploma	Action learning guilds	Certs, Evaluation, Re-entry, Closing		PARTY!
Sun 17/4							CLEARING UP & GOING HOME				

Fully outdoors session
Site visit (mainly outdoors)
Travel time
Free time

A permaculture design course timetable laid out in a spreadsheet

Now, a course runs to a timetable, but you might instead want to design a process that starts whenever inspiration strikes, such as writing music. This was something I set out to do as part of my Diploma portfolio. Again I wrote words on pieces of paper to help me organise the chain of events before drawing the final arrangement onto a ***flowchart***.

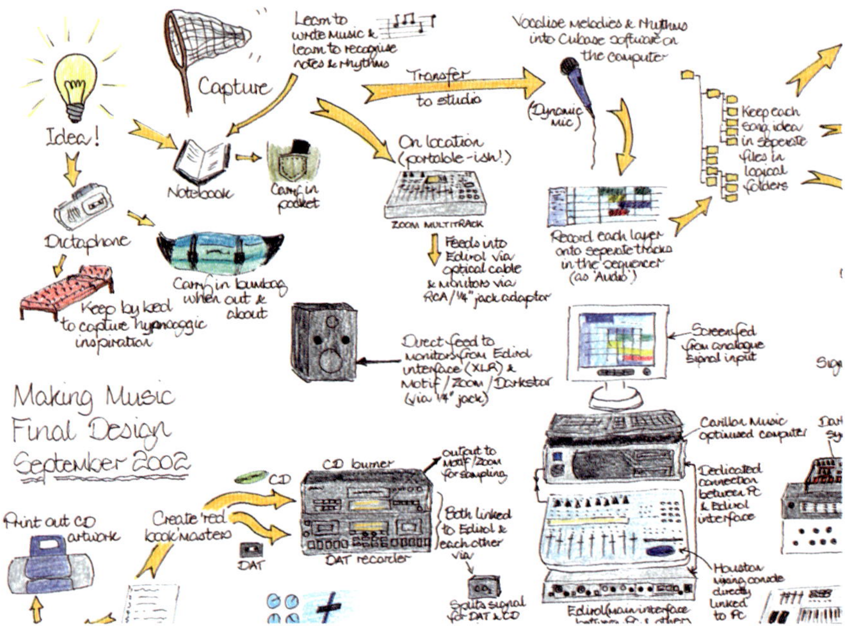

Detail from my making music design flowchart

Identifying optimum flows like this can often show you where best to physically place tools and resources. In this case that meant a notebook and pen in my pocket and a Dictaphone by the bed. I'd previously determined that the latter was a much more efficient way of recording my dreams than waking up to write them down as it allowed me to stay in the dream state and remember more details. In my studio I later used a ***micro-zoning exercise*** to lay out my recording gear efficiently. Don't be afraid to adapt existing tools to meet your own needs; I decided in this case that zone 1 was what I could reach while sitting at my desk, zone 2 what I had to stand up to reach, zone 3 what I had to stand up and turn around for, and zone 4 outside the room (though it could instead have been inside a cupboard or filing cabinet if I'd had room for either).

Mind maps can be a great way to present your ideas too. In another of my designs I considered how to get the best from my Diploma accreditation process.

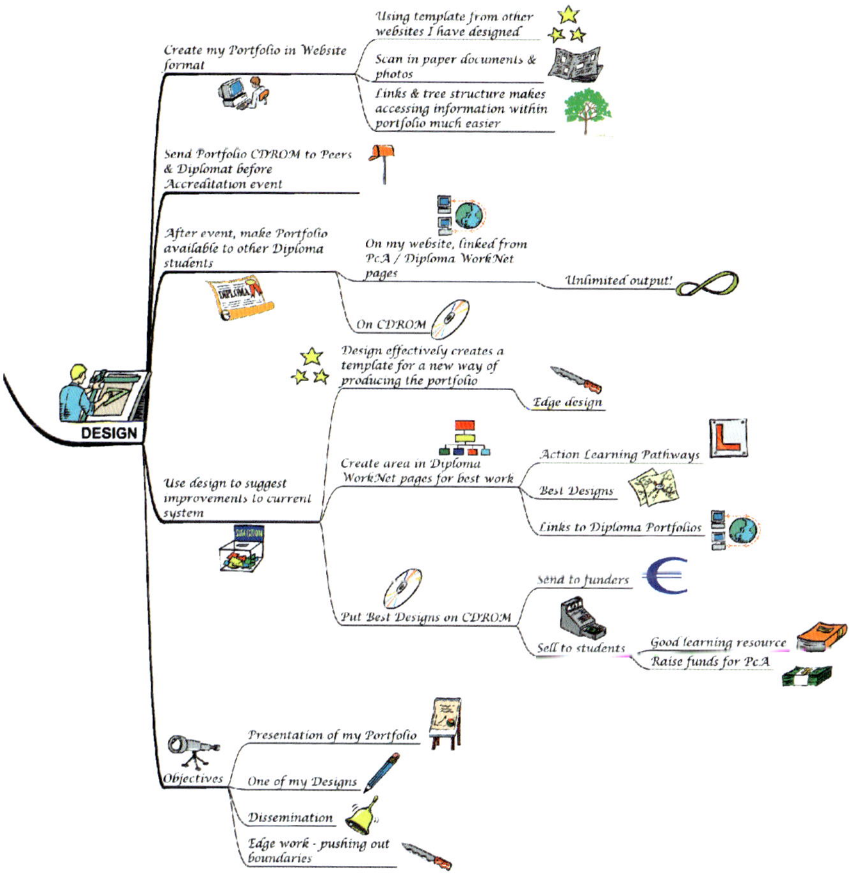

You don't need computer software like this to make them, though I find it provides extra flexibility in spreading ideas evenly across the page. This design was so complex that to make sense of it all on one page was difficult. As a result, I created an overview page and then a series of mini mind maps outlining the detail in each branch. Of course, flowcharts and mind maps often only show basic relationships between elements. To show timings we might again use a ***Gantt chart*** like the one I introduced under *Implementation*.

Designing zone 00
(creating a personal permaculture plan)

When designing such a plan for myself, I started off with the ***prime directive*** of permaculture:

> *"The only ethical decision is to take responsibility for our own existence and that of our children."*

To take responsibility is to become more conscious of all our actions. Whenever we choose something, we are voting for it being present in our world. That is why Gandhi suggested that we 'be the change'. Our thoughts determine our actions, so approaching any such design from an attitudinal perspective is a good start. Unwanted habits are just ***patterns*** that can be replaced, but we need to do so with more powerful ones for this to be successful. Deciding to give up a thing is not enough, we have to want something else with an even greater passion.

We are our own greatest asset, so when designing for ourselves one of our focuses should be on health; disregarding this will reduce our effectiveness at everything we set out to do. Much has already been written about the importance of diet so I won't repeat any of it here. However, the human body we all inhabit evolved over many millennia under specific conditions and needs more than just good food to remain healthy.

One key to effective design is ***dynamic balance***. Nature keeps moving, but wherever we go we find a dynamic stability; the river adjusts its course to bypass new obstacles, but still flows from the mountains to the sea. Movement is vital to life, as we can see when water stagnates. Likewise, our own bodies need to move in order to remain healthy, and yet our modern lives leave many of us inactive most of the time.

Crucially, Pete Egoscue[32] points out that not only do we need movement for health, but that inactivity *doesn't affect all body systems equally*. It's unlikely that any generation has ever spent so much of their time sat on chairs. This profoundly unnatural activity leads to weaker lower back and abdominal muscles, but strong hip flexors. This imbalance results in an exaggerated curvature of the spine which leads to pain and injury, but also affects our health in many other ways.

To follow the principle of ***working within nature***,† we need to learn the importance of how our bodies are biomechanically designed to work and what happens when we ignore this – an observation that Alexander teachers and the growing movement of barefoot runners have recognised. I find it interesting too that we see the same patterns in our human social systems. Imbalance leading to tension. Spinal misalignment or global inequalities – both lead to stress, inefficiency and pain.

One tool I use when teaching away from home is a ***diary***. I use it to record how well I sleep, what I eat, how much I exercise, how I'm feeling, what my energy levels are like. It allows me to notice patterns that can help me look after myself better. It was through using one that I discovered my intolerance for gluten.

Personal patterns~ phenological diary				Additional:	Wi-fi / DECT nearby?		Corridor lights?
					Room dowsed?		Clunking doors?
	Day:						
Weather	am: pm:						
Sleep	Curtains? Hours? Dreams? Undisturbed?						
Energy level?	Out of 10 ?						
Mental Physical (first thing) Emotional	☺ / ☹						
Breakfast	What? Quantity Quality						
Morning Activities	Physical? Mental? Etc.						
Energy level?	Out of 10 ?						
Mental Physical (morning) Emotional	☺ / ☹						
Lunch	What? Quantity Quality						
Energy level?	Out of 10 ?						
Afternoon Activities	Physical? Mental? Etc.						
Mental Physical (afternoon) Emotional	☺ / ☹						
Evening meal	What? Quantity Quality						

Noticing our own rhythms is a really important part of any self-design process. We're familiar with our daily cycles of sleeping and waking, but it's also been shown that our energy levels cycle throughout a day.[33] So going with the flow and having a snooze after lunch can actually help us to accomplish more.

† *An adaptation of the perhaps more familiar 'work with nature', but emphasising that we are a part of nature, not apart from it.*

Such a diary can help us to identify our best times for doing particular things; maybe you're a morning person and you've noticed that's your most creative time? We can design one to record both what's going on for us (our '***zones***') and also how we're affected by what's going on around us (***sectors***). Those sectors might include how the volume of traffic changes as we cycle to school or work at different times each day, allowing us to identify when it feels safest to travel. Or perhaps noticing the noisier times of day if quiet is important for what you wish to do. It may be that using the ***exclusion method*** you discover that your yoga zone is entirely defined by your children sector! As with any design, the observation stage is all-important, the great thing about designing our own lives being that each day we get the chance to try out something new.

Of course, a personal plan could address *everything* we do, so chunking it down into a collection of designs makes it easier to figure out. This is what I did when I designed my own life path as part of my Diploma portfolio. I divided it into sections covering areas such as food, water, exercise, shelter, energy, travel, relationships, communications, clothing and so on. There's a lot to consider, so go easy on yourself. Nature evolves slowly. ***Make small changes*** and reflect on their effectiveness. Gradual changes are also more likely to become permanent.

Our personal ***resources*** include our own skills and experience. What are we *already* good at? Don't stop at what your society currently values. Our life experiences have made us all experts at a diversity of things. How can we use those things to design the life we want to live? Our ***limitations*** too are gifts and can lead us to seek out important interdependencies with others. We are all an ecosystem within a larger ecosystem and cannot exist independently of nature or of each other.

A *right livelihood* involves providing something of value to the greater community of which we are a part. It wasn't so long ago that doing anything else would have meant starvation, as communities relied completely upon each other. We live in strange times now when many of us don't provide anything of real value anymore. Until quite recently in Britain, 60% of us worked on the land growing food. Now, through the use of machinery and fossil fuels that figure has fallen to around 1%.

When setting out to design a right livelihood for myself ten years ago, I considered the principle of redundancy: ***multiple elements for each important function***. As a result I created a poly-income for myself rather than only contributing to society in just one way, a vulnerable path that most of us now follow in employment. Many traditional jobs were dependent upon the seasonal availability of resources, which meant that for many having multiple ways of making a living wasn't just a good idea, it was essential.

I realised that I had to identify my future goals and work backwards from there. To get to where I wanted to be, what did I need to do before that? And to achieve that, what did I need to do before that? And so on. Then, in a self-as-client interview I asked myself what it was that I enjoyed doing most. Around that time I'd read that a reliable guide to being on the right track was to ***follow your excitement*** and to this day I'm glad that I did, or I wouldn't now be spending most of my working days enjoying a real sense of purpose. It might seem hard to imagine, from where you are now, a future where you are valued by your community for what you love doing, but I feel we all have important gifts to share. Like all good design, it's about identifying the place where these will be valued most. Life should be enjoyable and satisfying for us all, not just for a lucky few, and our communities would be so much better for it.

As well as asking ourselves what skills and experience we have, we might also consider those we'd like to acquire. What would make us a more valuable member of a less wasteful society? Following a learning journey of our own choosing can also restore lost self-confidence if we were devalued by a school system that only catered to certain learning styles. Of course we can all feel constrained by peer pressure and our cultural expectations too. These are powerful holding patterns, which originally evolved to protect community members from perceived threats. Respect them, but don't let them bind you.

Having physical tools and resources is also important, though there's a danger that if left unused we waste energy in maintaining them. From a patterning perpective, an efficient ***net*** harvests what it needs and allows through what it doesn't. If the latter is allowed to build up it eventually creates problems.

These days many of us are more likely to suffer from being *over* rather than *under*-resourced, the excessive clutter filling up otherwise useful space. This is why community structures like *Freecycle* are so good as they store useful things out in the wider network rather than in our valuable zone 0 spaces. Heaps of clutter require a bigger house to store them in which means more expense, higher insurance and so on, all of which have to be paid for, ultimately by trading our precious time.

Again, attitude is an important aspect of our personal plan. To *Freecycle* is to trust that when we do need them in the future, those resources will become available again in the network. Ironically we collectively need far fewer resources for this to work than if most of us put those things away 'in case we need them someday'. When we do this, many of them just slowly deteriorate unused. Often the best way to preserve a tool is to use it, which leads to it receiving much more attention in the form of maintenance than if it were tucked away in storage.

By way of a model, nature abounds with a form of trust, without which everything would grind to a halt. If a tree didn't shed it's leaves each autumn, many soil organisms would go without food, affecting the whole food web. The tree in turn would benefit less from the soil building services of these many creatures and life as a whole would be far less abundant. Wealth in a system comes not from the level of resources, but from the rate at which they flow around it.

Some might say that to trust enough to join in, those resources need to be already in circulation, but *someone* has to set up such projects where they don't yet exist.

See a gap? Maybe it could be your gift to the community...

Designing social structures

One of the keys to a truly sustainable future for us all will be how well we design the social structures around us, and as a permaculture designer you'll be one of the best qualified to help do this. To create successful social structures we need to consult across the community, which makes this a little more complex and time-consuming than any design for ourselves or just a few clients. Trying to consult so widely without a tried and tested process can lead to frustration and disinterest from all sides. So for the initial consultation stage, either bring in an external facilitator or get trained up yourself to step into this role. A successful consultation (observation phase) can make or break a new project, so it's worth investing in doing it well.

Any of the processes that I introduced previously: ***World Cafe***, ***Open Space***, ***Holistic Goal-setting***, ***Transition Culture***, plus others like ***Dragon Dreaming***, can be good tools for finding out what people want. You'll have plenty of great ideas of your own, but the first question should always be *does the community want this?*. These tools will help you to find out.

As part of this information gathering process, make sure you identify the skills and experience people have within the group. Ultimately, your ability to put the right people together in the best place, and to provide them with the tools they need, can make all the difference. ***Everything gardens***: identify who's already got a passion for a particular thing and give them a chance to be appreciated for it by the rest of the community. Malcolm Gladwell, in his book *The Tipping Point*, highlights the importance of key networkers to any new project's success. ***Use the least effort for the greatest effect***: identify these *connectors* (as he calls them) in your community and involve them in some way. In publicising a new project, it's not *how many* people you tell, but *who*.

While it's important to find key people to help move any new project forward, long-term resilience will require a diverse group of others in supporting roles. While a dynamic person can focus a project, their loss can cause its demise. Make sure that you have ***multiple people in each important role***.

Whilst interviewing the community could dominate surveying time, social designs will most likely also involve some non-living elements, like land and buildings. These places, where the people will interact, make up the landscape of your site survey. How suitable are these facilities for the purpose? Can they be improved? Are they in the best place? Getting the location of facilities right can make or break a project. Where do potential users live? In a garden share scheme, map homes with small or no gardens and also unkempt larger ones. Where do these people travel (***desire lines***)? Map ***zones*** and ***sectors*** to determine the best location for the community hub. If rent is an issue, look for other groups that could share the costs with yours. Map ***resources*** and skills in the area. Identify other local groups that might be good partners. Identify why this hasn't happened before – what are the key ***limiting factors***? Seek out success stories from other places where this has been done – do they share a useful pattern that could help you?

Social structures also rely on a sense of equality to bring new people in and build numbers to become resilient in the face of outside influences (***sectors***). These might include economic issues or changes in legislation, but a strong community will be able to support its members and respond accordingly. The flourishing ***Transition Network***, co-founded by permaculture designer Rob Hopkins, seeks to achieve this by offering local communities a pattern to develop resilience in the face of both climate change and peak oil.

Some projects, like the increasingly popular community garden, may start with a blank canvas. Such a case offers plenty of opportunities for ensuring a sense of ownership by involving everyone right through the process of design and implementation. Create something that people actually want and they'll be motivated to protect and enhance it.

One thing to be aware of – if you decide to step into the role of facilitating the group process, it will be harder for you to be both completely inclusive and also effectively share your land-based design knowledge. This is another reason why it's sometimes better to bring in a trained outside facilitator. Begin the process with a clear sense of fairness, and people will feel safer and valued within it. You'll get more from the group too.

In the end we all benefit from having strong networks around us. They are for instance an excellent way of saving and sharing valuable seeds and plants. I once created a garden where almost all the plants were been gifted to me in some form. Ironically, that gift was later taken away by a bull-dozer, but not before I had shared some of my own surplus with others. As well as protecting many important heritage varieties, seed saving networks are a great place to meet other gardeners and share experiences.

Seed packets beautifully decorated by Transition group members

Often the best way to get a new initiative started is to choose something that has a high chance of success and begin to build a network from there. Once established it will be easier to start subsequent projects and you'll be more experienced too. Make sure you identify all the levels of systems and subsystems that you are designing for. In my aikido design I soon realised that to optimise my own learning I also had to help meet the needs of my peers, my Sensei, the Club and the School as a whole.

Using permaculture to design in this way is I think still a hugely unexplored realm. How do the principles apply to our design? What about the ethics? If things aren't going as well as had been hoped for, where are the key bottlenecks in the system? Perhaps there are plenty of ideas, but no one putting them into practice? Where can we make small interventions that result in big improvements? ***See solutions rather than problems***; dealers may be trawling *Freecycle* sites to sell those things on, but don't lose sight of the purpose of the network – to keep resources out of landfill. From that perspective this is a success!

Most of all, remember to make plenty of time for celebration. These moments are the glue holding communities together. Celebrate the end of projects, but also at points along the way. Celebrate the passing of the seasons. Celebrate each others gifts...

Reflecting on the Process

No design is ever *completely* finished. You might walk away and leave it in someone else's hands, but whatever you set in motion will still demand their attention to some degree. And I should hope so too, because you will have set up energy and resource harvesting systems for them. No design will ever be perfect either; to achieve that would require the combined imaginations of an infinite number of permaculture designers. Even then, site conditions change and it would soon need tweaking to optimise it again. This would require not just the ongoing monitoring of its real world performance, but also reflection on what you observe.

Life has become so busy for many of us that we rarely find the time to give to this vital stage of the process. We're often told that we learn best from our mistakes, and that's a good thing if indeed we do. Mostly we go on making the same mistakes over and over again and do our best to sweep them under the carpet. I propose we start a culture of ***celebrating*** those ***mistakes***, to get excited about them and examine what it is we have learned from each experience. Then take that lesson into our next process and allow it to make us a better designer. Evolution has been so successful because life makes lots and lots of mistakes; trying out many things that don't work eventually leads to discovering some that do.

So, ask what went well about each process, but also what didn't go to plan. What areas of the design didn't perform as well as you expected in the real world? What have you learned from this? How will this make your next design even better? ***Have you had fun?*** I certainly have and continue to do every day. I've learned *heaps* from doing a lot of things wrongly. I hope these pages will save you repeating some of my errors, so you can share some new ones of your own later with others. From where I am now I see my own Diploma portfolio is full of holes. Hooray! I've learned something these last eight years. This guide is a reflection of my understanding right now. Please let me know your thoughts for the next revision and in the meantime celebrate my errors with me. I'm still learning too.

A Call to Action

As you may have noticed, I love the process of design. I love the creativity that emerges from within us when faced with constraints – and despite those big ones looming up ahead I'm really excited about the future. I'm excited because I keep meeting people with a great desire to make a positive difference in the World. People just like you.

Thank you for reading this book – I hope it has helped you to make more sense of the permaculture design process. Another principle: ***information is the critical potential resource***. By sharing much of what I've learned here I hope to support you in creating many excellent designs that will in turn inspire others to do the same, but *you have to do something with it*. So right here, right now, I'm handing you the baton...

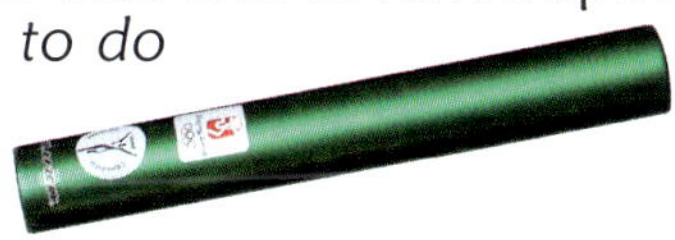

The good news is that there's already an expanding network of exciting and innovative permaculture projects being set up across Britain and beyond. Many you can visit for inspiration, one way being through the Permaculture Association (Britain)'s LAND demonstration network, which at the time of writing covers just England. LAND learning sites can all be visited by arrangement, though many of them provide tours rather than internships. If you've a need for more in-depth action learning, then WWOOF (World Wide Opportunties on Organic Farms) involves volunteering in exchange for food, accommodation and the chance to learn about organic farming.

Visiting such sites can be really inspiring, but why wait? Start now – at your back door, or windowsill if you have no garden. Do *something*. Small is just fine if that's all you currently have time for; Bill Mollison encourages us to ***work out from well managed areas***, so work within your capacity to increase your chance of success. Successes inspire further action. Failures make you wiser. *Either way you win*. One more thing, we've all got to start somewhere and I believe we're exactly where we need to be, to do what we're here to do.

Have fun on your journey and be an inspiration to others. ☺

Appendices

These final pages guide you to some sources of further information, including a few of my favourite permaculture related books and the link to download worksheets and mind maps, higher resolution versions of images and all of the design process flowcharts in printable form. I hope you find them useful.

References

1. *Permaculture: A Designers' Manual* (PDM) by Bill Mollison p78-79
2. A good primer is *Thinking in Systems* by Donella H. Meadows
3. *Introduction to Permaculture* by Bill Mollison, chapter 1
4. Colleen Stevenson's work can be seen at www.colleenstevensongraphics.com
5. *Biomimicry* by Janine Benyus
6. See example at Mark Fisher's excellent website, www.self-willed-land.org.uk Mark also offers a good description of the design principles there for such use.
7. *Six Thinking Hats* by Edward de Bono
8. *People & Permaculture* by Looby Macnamara
9. *Permaculture One* by Bill Mollison & David Holmgren
10. *Nature Diary* by Janet Marsh
11. *Changing Nature, eco notes of a digital woman* by Helen Moore
12. The Natures Calendar Survey, www.naturescalendar.org.uk
13. From the 'Bridging Industry to Permaculture' DVD – EcoSutra
14. *PDM*, Mollison p16
15. *The Permaculture Way* by Graham Bell p71-2
16. *PDM*, Mollison p15
17. *PDM*, Mollison p46-7
18. *The One-Straw Revolution* by Masanobu Fukuoka p33-4
19. *PDM*, Mollison p37-9
20. *PDM*, Mollison p47-8
21. *PDM*, Mollison p13
22. *PDM*, Mollison p70
23. *A Pattern Language* by Christopher Alexander et al
24. Fuzzy (non-binary) logic is the study of such things.
25. "Strip cropping wheat and alfalfa to improve the biological control of the wheat aphid *Macrosiphum avenae*..." – Ke-Zheng Ma et al
26. *PDM*, Mollison p60
27. *PDM*, Mollison p62-4
28. "A map is a visual representation of an area – a symbolic depiction highlighting relationships between elements of that space such as objects, regions, and themes." Wikipaedia
29. Darren Doherty does this really well in his 'Keyline on the Beach' videos.
30. *Water for Every Farm* by P.A.Yeomans
31. Available to view at www.aranyagardens.co.uk
32. *The Egoscue Method of Health Through Motion* by Pete Egoscue
33. *The Power of Full Engagement* by Jim Loehr & Tony Schwartz

Full design process flowchart

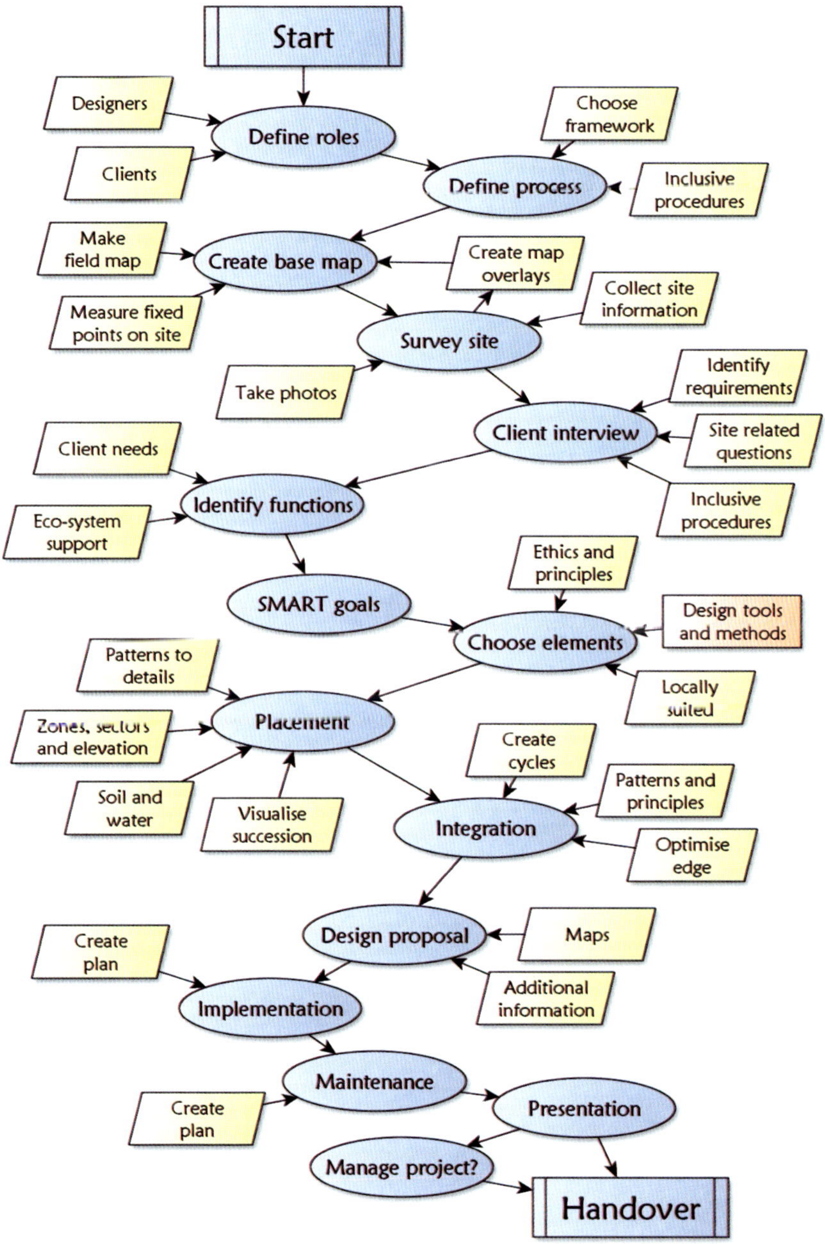

Resources

Online materials

The following supporting materials can be downloaded from:
www.learnpermaculture.com/index.php/design-guide-downloads

- All my own drawings at higher resolution
- Design worksheets and tables
- All process flowcharts
- Mind maps (limiting factors/resources/microclimates)
- Soil testing charts
- and maybe even more...

Recommended reading

A by no-means exhaustive list, these are just a few of my favourites:

Permaculture Design

Introductory:
Beginner's Guide to Permaculture – Graham Burnett
Permaculture in a Nutshell – Patrick Whitefield

Intermediate:
An Introduction to Permaculture – Bill Mollison
Permaculture One and *Permaculture Two* – Bill Mollison / David Holmgren
The Permaculture Way – Graham Bell
The Basics of Permaculture Design – Ross Mars

Comprehensive:
Permaculture: A Designers' Manual – Bill Mollison
Permaculture: Principles & Pathways Beyond Sustainability – David Holmgren
The Earth Care Manual – Patrick Whitefield
Earth User's Guide to Permaculture – Rosemary Morrow

Permaculture Gardens / Edible Landscaping

Designing and Maintaining your Edible Landscape Naturally – Robert Kourik
Gaia's Garden – Toby Hemenway
Sepp Holzer's Permaculture – Sepp Holzer
The Permaculture Garden – Graham Bell
The Permaculture Handbook: Garden Farming for Town & Country – Peter Bane
The Permaculture Home Garden – Linda Woodrow

Systems and Patterns

A Beginner's Guide to Constructing the Universe – Michael S. Schneider
Heaven and Earth – ed. Katherine Roucoux, Phaidon
Hidden Nature – Alick Bartholomew
Nature: the Mother of Invention – Felix Paturi
Thinking in Systems – Donella H. Meadows

Soil and Water

Farming in Nature's Image – Judith D. Soule & Jon K. Piper
The Humanure Handbook – Joseph Jenkins
Mycellium Running – Paul Stamets
The One-Straw Revolution and *The Natural Way of Farming* – Masanobu Fukuoka
Rainwater Harvesting for Drylands and Beyond (volumes 1 & 2) – Brad Lancaster
Teaming with Microbes – Jeff Lowenfels and Wayne Lewis
The Water Book – Judith Thornton
Water for Every Farm – P. A. Yeomans

Trees

Creating a Forest Garden – Martin Crawford
Edible Forest Gardens (volumes 1 and 2) – Dave Jacke
How to Make a Forest Garden – Patrick Whitefield
Sowing the Seeds of Change – Treesponsibility
The Woodland Way – Ben Law

Food

Local Food – Tamzin Pinkerton & Rob Hopkins
Organic Gardening and *Salads for All Seasons* – Charles Dowding
Perennial Vegetables – Eric Toensmeier
Plants for a Future – Ken Fern
The Winter Harvest Handbook – Elliot Coleman

Buildings

A Pattern Language – Christopher Alexander et al
Shelter and *Home Work* – Lloyd Kahn
Spirit and Place – Christopher Day

Urban / Communities

The Abundance Handbook – Grow Sheffield
Sacred Economics – Charles Eisenstein
Toolbox for Sustainable City Living – Scott Kellogg and Stacey Pettigrew
The Transition Handbook and *Transition Companion* – Rob Hopkins

Personal

Coming Back to Life – Joanna Macy
The Earth Path – Starhawk
Eat More Raw – Steve Charter
Find Your Power – Chris Johnstone
Finding Earth, Finding Soul – Tim Macartney
Barefoot Running: Step by Step – Ken Bob Saxton

Miscellaneous

Biomimicry – Janine Benyus
The Buzz About Bees – Jürgen Tautz
Left in the Dark – Tony Wright
Nature's Operating Instructions – Kenny Ausubel with J. P. Harpignies
Sustainability, Consciousness and Climate Change – Steve Charter

Glossary

Agroforestry – the combining of trees with agricultural or horticultural crops, animals, or aquaculture for the many extra benefits that trees provide.

Aspect – the direction a slope faces, primarily in relation to the sun.

Bunyip – a water level consisting of a transparent pipe, each end being attached to a vertical pole marked off in height increments. The pipe is filled with water to a convenient height half way up each pole (i.e. 1 metre for a 2 metre pole). The level is used in a landscape to identify either points of equal height for marking contours, or to measure the difference in altitude between two or more site elements.

CAD (Computer Aided Design) – the use of computer technology for the process of creating designs and design-documentation.

Capping – a surface crust or cap caused by heavy rainfall on bare soils.

Element – (from Systems Theory) this refers to an individual object, which when in interaction with other elements make up a system.

Forest garden – a food production and land management system based on woodland ecosystems, but substituting trees (such as fruit or nut trees), bushes, shrubs, herbs and vegetables which have yields directly useful to humans.

Function – (from Systems Theory) this describes the behaviour or purpose of a system or individual element. From our design perspective this is what we are wanting to achieve, rather than how we will do so.

Gabion – often a cage, cylinder, or box filled with rocks, soil or sand and used as part of an engineering structure; in our case most likely for erosion control.

Green manure – a type of cover crop grown primarily to add nutrients and organic matter to the soil. These are often nitrogen fixing species.

Guild – a mutually beneficial grouping of organisms: plants, animals, fungi, bacteria etc. around a central element, which is often the main focus of benefit.

Keyline® – a collection of design principles, techniques and systems for the development of rural and urban landscapes.

Keypoint – a point in a valley floor where water first slows and deposited materials begin to accumulate. The contour on which it sits is the keyline.

Mycorrhiza – fungi that form beneficial relationships with living plants, exchanging amongst other things, useful bio-available compounds for plant sugars.

Net and pan – a network of planting pits joined by channels, used on dryland slopes to direct run off water and organic matter to plants and trees.

Pattern language – a structured method of describing good design practices within a field of expertise, often architecture and landscape architecture.

Swale – a water harvesting ditch created on contour, used to slow run off and encourage infiltration, for watering associated trees and recharging the water table.

System – (from Systems Theory) is a made up of a collection of elements or sub-systems interacting with each other. Their overall behaviour can be determined by that of individual elements and/or the result of their relationships, often through feedback loops. Systems have functions, or purposes, by which they may be named.

Thermosyphon – a method of passive heat exchange based on natural convection which circulates liquid without the necessity of a mechanical pump.

About the Author

Aranya has been experimenting with what permaculture can do since his design course epiphany in 1996. In the years that followed he designed a collection of gardens, along with a few other non-land based designs, writing them all up to gain his Diploma in Applied Permaculture Design 2003. Since that time he's followed a teaching pathway, that now involves delivering on average ten two-week design courses a year. Along the way he's met many amazing people that give him hope for the future and who in no small way have influenced the content of this guide. Aranya is also a long-term barefooter, certain that this helps him be more aware of nature, stay grounded and walk lightly on the Earth.

Websites

Aranya's blog: ***www.abundantfuture.co.uk***
Aranya's main website: ***www.aranyagardens.co.uk***
Course listings: ***www.designedvisions.com***
Online learning: ***www.learnpermaculture.co.uk***

Check out these websites for more information and support:

Permaculture Association (Britain): ***www.permaculture.org.uk***
Permaculture magazine: ***www.permaculture.co.uk***
Permaculture Global Network: ***www.permacultureglobal.com***
Permaculture Research Institute: ***www.permaculture.org.au***
Permaculture Activist (USA): ***www.permacultureactivist.net***
TED (many inspiring talks): ***www.ted.com***

Credits

All photos and images © Aranya except: Aranya's photos of design course students work – p141-144,147-8,155 / Crap Cycle Lanes (Eye Books) – p59 / istockphoto – ps x,xi,2,4,7,10,30,46,47,56,80,92,112,138,152,162,166,172,183,185,186 / Colleen Stevenson – p14 / www.magneticdeclination.com website – p49 / Desbois design – p64,69 / Darren Doherty – p145

Solutions for Changemakers

Every one of us who picks up a seed packet, upcycles a piece of furniture or invests time and energy in a community project is a changemaker. The minute we start to make small but incremental changes in our lives, we become agents for positive change and the ripples of our influence spread out around us.

Permanent Publications produce a range of books to empower and inspire changemakers the world over, from no dig organic growing, food forests and permaculture design, to natural building, renewable technology and connecting with nature.

If you enjoyed *Permaculture Design*, why not try these titles:

PERMACULTURE DESIGN COMPANION
Jasmine Dale
£19.95

THE EARTH CARE MANUAL
Patrick Whitefield
£45.00

PEOPLE AND PERMACULTURE
Looby Macnamara
£21.00

For the full list of our solution-based titles visit
www.permanentpublications.co.uk

Our books are also available in North America and Australia:
www.chelseagreen.com/publisher/permanent-publications

https://peribo.com.au/product-category/titles-by-publisher/permanent-publications

permanent publications
EARTH CARE, PEOPLE CARE, FUTURE CARE

Permanent Publications also publishes *Permaculture* magazine